# ENTROPY RULES

## The Myth of Human Exceptionalism

GRANT GOODBRAND

Library and Archives Canada Cataloguing in Publication

Paperback: ISBN 978-1-0689397-0-9
Ebook: ISBN 978-1-0689397-1-6

Cover design: John van der Woude | jvdwdesigns.com
Interior typeset: Sammy Chin

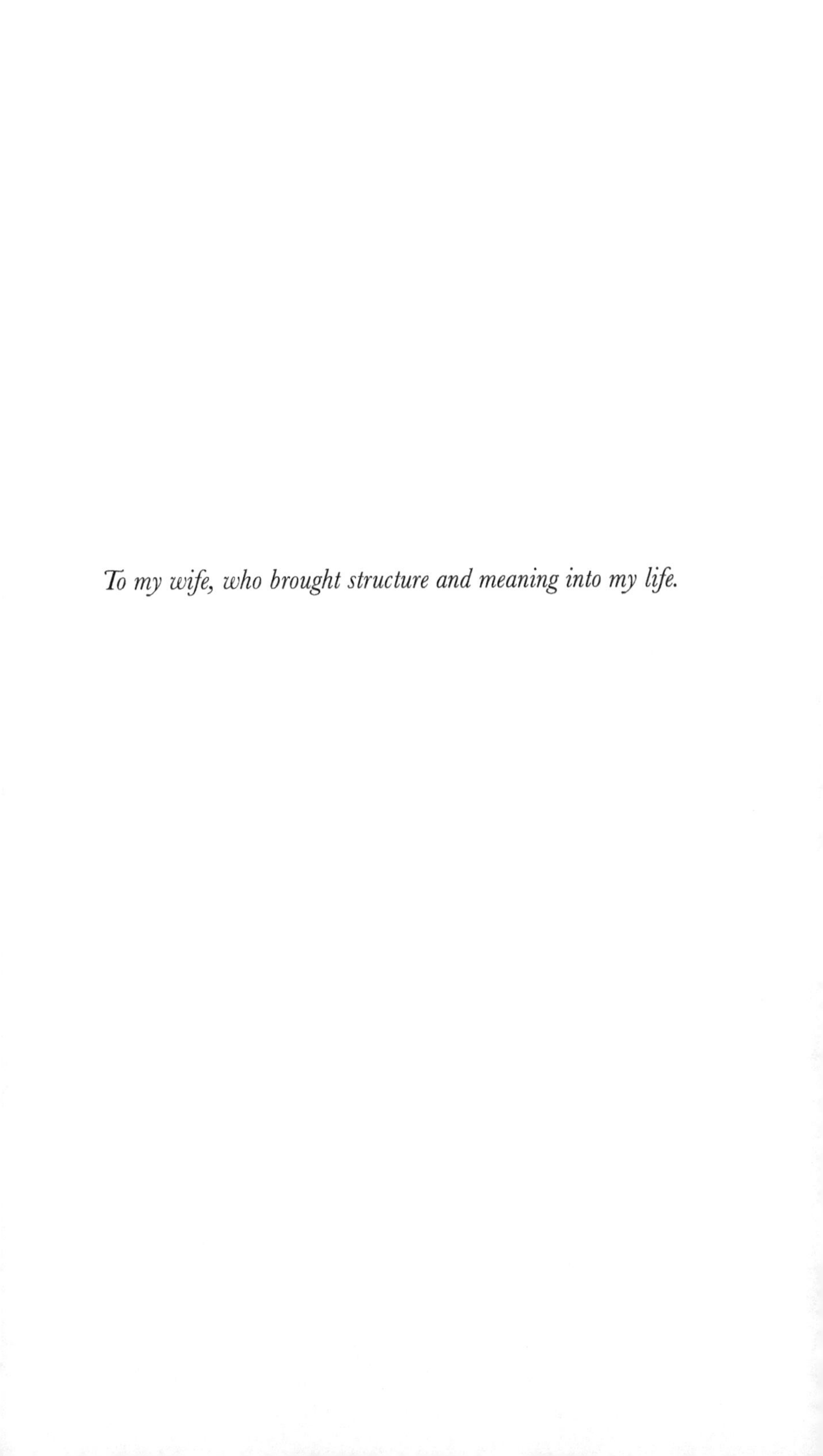

*To my wife, who brought structure and meaning into my life.*

*In three words I can sum up everything*
*I've learned about life:*
*It goes on.*

— Robert Frost




*"All that is very well,"* answered Candide,
*"but let us cultivate our garden."*

— Voltaire

# Contents

# Preface

The questions informing the therapy of the people I worked with for over fifty years were: "Why do I do what I don't want to and don't do what I want to?" "I'm in pain and I don't know why." "I don't know what to do." "My relationships aren't good." The therapeutic answer to these questions was a form of self-examination to determine a coherent personal narrative inclusive of all their conflicting experiences and emotions.

I was attracted to psychotherapy because I alternately experienced turbulence or a vacuum within myself, as well as remoteness from others. I traced these painful disruptions' origins to my first years of life, which were spent in foster homes, and the violent conflicts within my adoptive family. My bias in working with my patients, and making sense of my own pain, was to look for meaning in the beginnings of processes and to favor objectivity. Practicing psychodynamic-based and psychoanalytic-based therapy for fifty-four years reinforced these conditioned biases.

Now, in this book, I argue that the inescapable laws of nature and the universe are the underlying causes of our falling apart into compulsions, pain, confusion, destructiveness, and self-destruction. This is a story — a reflection based on science, which posits that mathematical equations are the language of our reality, not words. This book is an experiment in thinking about our earthly presence from outside our human point of view.

The search for a therapeutic for the confusion and painfulness of the human condition has been sought many times in philosophy and religion. Plato was the first to attempt a corrective for the human species' built-in distortions of its world and of itself. Humans are chained to pale, schematic representations of reality. Congratulating ourselves on our abilities to finely differentiate among ephemera, and championing our shallow reasoning and beliefs, we refuse to directly examine the causes and meanings of existence. We are chained prisoners that prefer stories to knowledge, and stasis to change. Plato prescribes a radically different point of view that seeks the sources for our existence in universal truths that free us from our crippled, self-punitive state. The German philosopher Immanuel Kant argued that our species' reasoning and experience are contained within categories built into us as an umwelt. What things are in themselves — what he called the noumenon — is unknowable. On the other hand, he thought that the general categories allowed for predictability and that categorical imperatives could apply universally to the moral behavior of all members of the human species. Recognition of these truths could ameliorate the ills of the species. The early twentieth-century philosopher Ludwig Wittgenstein cautioned us that the limitations of our language restrict our experience and our knowledge. Our best therapy was to define to the best of our ability the meanings of our verbal interactions and not to try to force our concepts to describe what is not part of our everyday experience. I propose that the answers to why the human condition is so painful lie within science. Scientific knowledge should be applied to every aspect of our culture and behavior.

As individuals we make up just-so stories that mask our ignorance of why we act the way we do. Our tactics and generalizations are based on slight experiences. In psychotherapy, we attempt to formulate a coherent narrative from an intense examination of our motives and whatever history is available. We explore fallacies and rigid constructs that undermine our effectiveness. Our complaint is that that which we want to do and want to do with all our heart we do not seem to be able to do and that which we don't want to do and do not want to do with all our heart that we do all the time. The therapeutic hope is that with careful and evidence-based appraisal of our motives, history, and fixed beliefs, we may modify our destructive and self-destructive trajectories that subvert our intentions and betray our hopes.

My search for our species' therapeutic transcends the search within the personal narrative. It is rooted deeply within my anger and aversion for the lies, pain, cruelty, and greed that characterize the greater part of our species' experience.

In the early 1950s, during my teens, I concluded that a nuclear war, which would alter our civilization and potentially end our species, was unavoidable. I could not imagine that human beings could desist from using any available weapon. During the Suez War in 1956, when such an event seemed imminent, I prayed to be given five years to figure out why we existed. It seemed awful to die without knowing.

I look across and down a large valley at my seven- and eighteen-year-old selves while presently standing on an escarpment of experiences. I barely recognize a fraction of their existence, but I share the questions that stirred their disquiet. Their unease and questioning have been a driving force throughout my life.

# Introduction

The ancient Greeks said that those the gods wanted to destroy they first made mad. By madness they meant hubris — narcissistic extravagance. Human exceptionalism is the conviction that we are not governed by natural laws that apply to the universe. We are the sole exception in nature, therefore superior to everything else. It is our presence that gives the universe meaning.

The belief in human exceptionalism exemplifies our power-driven essence. The myth making creates mythos-driven tribal and national societies and justifies a ruthlessly nature-sacking species. Bewitched by hubris, we privilege racism, nationalism, hierarchy, nepotism, female inferiority, inequality, greed, cruelty, war, the sacking of the planet, and unlimited consumerism. In this century, we are acutely aware of the cruelty and killing capacities we unleash on other species and each other. Some have compared us to a virus in the body of the planet. Destructiveness is at the core of our species' identity.

Ultimately, we have come to believe that we are part of a supernatural realm. We see ourselves as the paragon, focus, endpoint, and meaning of all that exists. We refuse diminishment to a small portion of nature that contributes to only one of nature's myriad processes. By believing there is a radical break between nature and us, we are trapped in a myth. Lost in human subjectivity, we have constructed an unreal world in our relationships, institutions, and belief systems. The result is detrimental to our lives individually and collectively, and to the planet. We can see this most clearly in our stories, which deny our fragility and comfort us by grandiosely claiming our superiority to everything on our planet, even to the universe.

A significant portion of humanity continues to believe that an eternal soul exists apart from nature and that a God created the world for our use. Even texts on neuroscience, biology, and physics use the concept of emergence to salvage Homo sapiens' superiority. The claim is that the mind is separate from nature and that humans have a particular autonomy and ontology. We want to weasel out of our total embeddedness in natural processes to maintain our privileged state.

We give primacy to phenomenology, which considers human subjective experience derived from our sense experiences as the basic data from which all truth is derived. This is a primary example of the arrogance of human exceptionalism, which makes the experience of one species the gold standard for all truth — "Man is the measure of all things."

Our self-importance is understandable because of our narrow point of view — our umwelt. Every cellular creature views the world entirely from its place in its environment, accessed through its sensory apparatus. The evolutionary niche into which we evolved shapes the meaning we give to the human story. Our short life, limited senses, and evolved conditionings distort our concepts and perceptions. Our belief system is the creation of a small animal on a tiny planet in a virtually infinite universe.

It is science alone that allows us to see ourselves objectively from the outside. With difficulty, we can place ourselves in alien scales of space, among fundamental and unfamiliar forces, within different mental organizations including ancestral forms of life, and within emotions and actions that are abhorrent and inhuman.

We are entirely embedded within matter. We do not stand apart. Our values and self-reflections embodied in stories that separate us from nature should be ruthlessly scrapped as we look at ourselves through the lens of science, particularly physics. Through science, we see ourselves from beyond the tunneled vision of our species. All planetary life is an insignificant part of a cosmic process, which has a well-defined plot, purpose, meaning, and denouement. The real story is not human-centered.

The incomparable predictive accuracy of the laws of quantum mechanics describes the fundamental reality in which we live. Forces and particles make up every part of us. With the insights of science, the door closes on our prior convictions and speculations about who and what we are. They become antiquated interests. Our privileged identity and purpose are undermined — evolution, neuroscience, and physics shatter our illusions of permanence.

Cognitively, science reshapes us. We, and all other cellular creatures, exist only in the molecular world. Our limited boundaries of space and time have been broken. Science has redefined what it means to be human and closes the large gap between and us and other animals, let alone other living things, such as plants and trees. The scientific project is now destabilizing societies, religions, and politics. It is only by accepting our real place within nature, described by physics, that we can confront the myths that distort our lives and create chaos.

We require an alteration in our point of view if we are to find a modicum of safety. By relocating ourselves to the scientific reality of our existence, we can achieve safety and give real meaning to our lives. Viewing ourselves enclosed in nature can be a container for our hopes. This change in viewpoint can change our reality, our moral presence, and the institutions we establish to create safety for us, our families, and our neighbors.

A science-based view of human nature and its meaning is truthful and therefore, paradoxically, more human than our just-so stories. Viewing ourselves from a cosmic point of view may lead us to safety within our limits. We must live within the bounds of natural law, away from the self-serving exceptionalism that leads us to cruelty and destructiveness.

In this book, I attempt to derive ethical values from the laws of physics and, by that means, offer an alternative within the limits of natural law to the destructive activities of humans. It is written to counter our central human claims of what is significant. Science places life (including us) as sub-entities within a much larger picture of the process of the universe, which seeks thermal equilibrium and the lowest energy state. This process requires the breakdown of composite matter. All cellular reproductive creatures, including humans, contribute to it. This process is key and governs and gives meaning to all our human activities.

Within the chapters, I make the following claims:

- The process of the cosmos is the breakdown of matter to release all relativistic energy, which will be devoted to the expansion of the universe. Temperature and energy differences will be resolved to the lowest energy state at a temperature near absolute zero. Humans have significance only as participants in this process.

- The sovereign and certain law of nature governing the breakdown of matter is the law of entropy — the second law of thermodynamics — as it applies to complex objects. It is also described by statistical mechanics applied to single particles. The basic process of the cosmos is the inevitable breakdown of all complex objects into simpler disordered particles and then into energy.

- Life is a chemical process to break down matter and eliminate gradients. This is the purpose of life, and because humans are a species of living things, this is our purpose.

• Along with energy and mass, the other fundamental is information — the laws by which anything and everything relates and how complex entities are created. Algorithms are the basic language on a universal level and are fundamental for the assembly of the human, among all other entities from the most basic to the most complex. This includes the organization of all living things. The script is mathematics.

• On our local species' level, the basic language is verbal, describing the relationships of the individual to what it senses and its relationship to others of its kind, individually and in groups. The script is written. Scientific language is precise, while verbal communication is ambiguous. With the continuing development of artificial intelligence, the scientific language will prevail in the description and direction of the species' life and the verbal will be marginalized.

• Life is the energy of the electron that is released in the process of the separation of molecules. Life is the movement of electrons. Life is matter, not spirit.

• The neuronal complexity of the human brain is more than adequate to process the chemical and photon bombardment of the brain by the environment and to arrive, by repeated experiences linked by neurons, at motor responses that distinguish between what is known and predicted and what is not. Culture is a brain-based process.

• Behavior is conserved from the single-celled to the multicelled animals. Behavior we consider human has existed from the earliest times of single-celled animals. It has developed evolutionarily ever since. We repeat.

• Humans evolved physically from the single cellular organism through the mammalian rodent and the offshoot primate. It is salutary to think of humans descended from the first mammals, the rodents, and to characterize humans as highly developed rodent offshoots. It is no wonder that psychologists use mice to determine the characteristic behavior of mammals, including humans, and study the brains of mice, which on a granular level are indistinguishable from the human brain.

• If we see our human behavior within the context of mammalian behavior and realize that we have, over time, modified the inherited behavior of other mammals, then the door is open to further modifications, particularly with biosynthesis, to attempt to eliminate the many destructive behaviors that bedevil our species.

• Our self-centered aggrandizement has caused us to exaggerate the resources that are available to us on this Earth as we assert our dominion over it and presume a limitless inheritance. We live in a delusion about the extent of our resources and our capabilities. This has led our species into danger.

• A general theory of morals contains three parts, all of them offshoots of entropy. Morality is first a consequence of the laws of physics describing the alternations between potential and kinetic energy. I suggest that these physical processes underlie the social organizations and conflicts within our species. These physical processes — manifestations of the law of entropy — drive human behavior.

• Humans acting as algorithms is central to the general theory of morals. Algorithms seek to replace other algorithms when experience teaches that one is more efficient than the other. We are algorithmic. This is a primary contributor to the defining characteristic of humans as cruel and aggressively destructive.

• We turn the liquidity of our brain's connections into solidity. This response to entropy is also part of the general theory of morality. The brain mechanically creates sums of sums of connections, at the highest level creating what we term "the self." The neuronal connections are constantly being remade, and therefore so are we. We live in a universe of entropy, and we fear our alteration and dissolution. Therefore, we try to fix, using memes and ideologies, "truths" that are just-so stories to give a permanent definition of ourselves individually and collectively. These constructions form limitations and become the platform for destructive behavior.

- Our species may be able to overcome its defining characteristics of greed and cruelty to become less self-destructive. I propose the goal, and some means for achieving Safety for Everyone. This will depend upon the internalization of the subject of this book, which is the renunciation of our hubris and acceptance of our very limited role dictated by the unfolding process of the cosmos.

Human life is difficult because the second law of thermodynamics prescribes the breakdown of the potential energy in complex matter into radiation and of the inexorable transition of organized particles capable of work into random particles that can never again be gathered into complex matter. Creating waste heat, waste matter, and disorder is a mandate written into our genetic code.

Is human civilization, therefore, a superficial layer of froth? To believe this would be very depressing. Our existence, functions, and behaviors are a minor part of the resolution of temperature and density gradients, and a correction to the loss of symmetry that created the complexity of matter. Viewing ourselves enclosed in nature can be a container for our hopes that changes our viewpoint, creating the safety that is presently lacking for our species and for individuals.

# PART ONE

## Why Are We Deluded? What Is Real?

Part One derives our myth of human exceptionalism from our separation from nature. It focuses on how we are part of nature, controlled by natural laws. The existence of life and our portion of it is derived from the natural processes that remove gradients and break down matter to return the universe to a grounded energy state of symmetry and unlimited expansion.

# Chapter 1

## Entropy: Everything Falls Apart

It sometimes seems to me that most of life is maintenance and replacement. I am almost always fixing something that is falling apart knowing that it will eventually fall apart again. It can be the house, finding food or finding clothing, finding files, doing taxes, or cleaning, or car repair, again and again. Then there are worries about whether the economy will hold or fail, or leadership will hold or fail, or relationships will endure or not. Communication is there, and then it is not as energy runs down. Then again there is the increasing repair to the body, which breaks down as one grows older. And then there is death. All of these seem like burdens, insults, injuries, and sometimes tragedy. But here is the truth: *This is the way it is supposed to be. This is physics, and this is the plan of the universe.*

### BREAKING DOWN GRADIENTS

Entropy is the process that drives the breakdown of gradients to complete the arc of the cosmos, restoring it to seamless energy with the elimination of matter. Entropy also creates the arrow of time, directing it forward and preventing a return to the past. In its initial state and as it developed the cosmos was homogeneous and isotropic. Everything was smooth; nothing stood out. Even particles were evenly distributed. Probably because of quantum fluctuations, there appeared minute fluctuations in heat and particle concentrations. As the universe expanded these tiny fluctuations became more pronounced, forming gradients in heat and matter. The concentrations of hot spots, associated with matter being compressed, increased at the same time as the spreading out of the radiation was lowering the temperature of the universe. So, now we have hot spots like suns and galaxies while the overall temperature of the universe is just 2.73 degrees kelvin above absolute zero. We also have high concentrations of matter, once again in suns, galaxies, and most of all in black holes and neutron stars, at the same time as we have a universe which is spreading out, creating a larger and larger void between the galaxies. The process of the universe is to eliminate these gradients and restore it to a uniform state of low energy and a constant temperature of just above absolute zero, with virtually no matter. All matter restored to being energy will expand, presumably indefinitely.

## LAWS OF THERMODYNAMICS

Thermodynamics describes the science of the relationship between heat and energy. Physicists have determined four Laws of Thermodynamics that govern this relationship.

The first law of thermodynamics, which anyone who took a physics class in school will remember, states that energy can be neither created nor destroyed but is transformed. The second law of thermodynamics states that entropy always increases. In a closed system, such as the universe, order declines to disorder over time. The third law states that as the universe approaches absolute zero, zero entropy is the equivalent to a perfect crystalline solid. (The fourth we can skip.)

It is the second law of thermodynamics that forms the foundation of all that follows in this book. Indeed, as the astrophysicist Arthur Eddington famously said, "The law that entropy always increases holds, I think, the supreme position among the laws of Nature . . . If your theory is found to be against the Second Law of Thermodynamics I can give you no hope; there is nothing for it to collapse in deepest humiliation."

The second law of thermodynamics states that heat will always go from hotter to colder and never the reverse. This applies to our bodies, to heat emanating from the Earth's molten core, and to the stars' heat that spreads into interstellar space, where the temperature is close to absolute zero. All coherent bodies in the universe — us, the stars — are ceasing to function and evaporating. Everything is burning up.

Close to home, the human body temperature is 37 degrees centigrade, while the average surface temperature on Earth is about 14.9 degrees centigrade. We require highly focused movement to keep us alive. We must constantly replenish our energy to maintain a sufficient body temperature and counter our constant dissolution. Even so, 80 percent of our living process is released as heat in respiration, radiation, and waste. Waste is defined as the inability to create coherence with many-bodied matter and push objects around to sustain work.

The dissipation of heat from our bodies is illustrated in the example of 300 people in an auditorium during a conference. Their bodies break down matter and salvage the energy of electrons while billions of trillions of particles spread beyond their bodies and, moving faster, interact violently, increasing the temperature of the room and even of the solid objects in the room. When they leave the room, the particles leave with them to spread out into the environment and cease interacting, becoming colder and colder. It is the random submicroscopic motions and vibrations of the particle constituents of matter that create the internal energy of a substance. This energy-movement is released with the breakdown of matter. Some of that energy may be directed, as when the ATP molecule in the muscle devolves, releasing an electron that assists in the muscle's contraction, which in turn picks up the spoon. Most of the motion's energy is dissipated into the environment as heat in random movement, never to be used again as part of a coherent piece of matter.

Farther from home, matter is also burning up, converting to radiation. Between 100 billion and 200 billion galaxies, with an average of a billion stars each, are burning and coming into equilibrium with the surrounding void at a temperature of 2.7 kelvins. Our star is at 5,778 degrees centigrade. Earth's interior, at 5,200 degrees centigrade, is liquid magma that swishes around an iron core and gives off radiation to the crust at 1,000 degrees centigrade. The radiated energy continues into the void.

To understand how this works, we must first understand heat energy. Within complex objects, those made of many particles, heat is latent because the object is inert — or in the case of particles and molecules, it is constrained. Work is the movement of an object or particle that moves another object or particle. However, in the release of their latent, controlled energy, objects lose some coherence. Some energy moves other matter, but some, frequently most, is released as heat into a cooler environment, where particles are farther apart and not interacting to create heat. The movements of particles become random; energy is transferred or transformed. Complex objects fall apart, and their energy will never regain coherence to be useful for work. The energy, formerly confined in matter, now becomes waste. (From our point of view, what is not work is waste.)

In contrast to thermodynamics, which describes the macroscopic movement of heat in objects and gases, statistical mechanics describes entropy by the movement of particles at the quantum level. Physicist Ludwig Boltzmann, in 1876, proved that microscopic particles have vastly more possibilities of being in a disordered state compared to the minuscule possibility that they can come together into an ordered state. Therefore, they inevitably become more disordered as they interact. Entropy is the inevitable process by which all isolated systems become disordered.

For example, if one takes a fresh deck of ordered cards and throws it into the air, the chances of those cards returning to the order they were in is negligible. Similarly, a burned-up object, reduced to its parts, will never reconstitute itself as a coherent whole. It can never be used to move other objects and create work.

Because the micro-states that create a macro-state are very few and therefore have a very low probability of existing, the macro-state will move toward a system with many possibilities.

Another way of expressing entropy is that every system (object) tends to achieve the lowest energy state, shedding the latent energy of its constrained movements. The cosmos seeks the elimination of gradients. Because it shares its energy with something else, and that something else has so many degrees of freedom, energy is divided into indiscernibly small portions. The vibrations or rotations of atoms and molecules transfer energy. Randomness applies to position, energy, velocity, and spin.

When an excited particle bounces into an unexcited particle, the low-energy particle increases in energy and the high-energy particle decreases in energy. Hence, there is a high probability that the energy will become evenly distributed rather than concentrated in a small part of the system. The subsystem phenomenon moving toward a lower energy state is called an instability. This is the process behind the second law of thermodynamics. The enormous potential energy stored in any macro-object as constrained motion at the level of molecules, atoms, and electrons is transformed into palpable motion and dissipated heat.

The second law of thermodynamics states that energy leads to heat diffusion, cessation of useful motion, and exhaustion of potential energy throughout the material universe. Potential energy is transformed into palpable motion and heat, which is then dissipated. Everything put together comes apart. The universe's destiny is the breakdown of all complex, coherent, low-entropy matter into incoherent, random-movement, high-entropy particles.

Physicist Hermann von Helmholtz, in the nineteenth century, concluded that all energy in the universe would eventually be transformed into heat at a uniform temperature. Natural processes will cease because there will be no place lower to which heat can flow; entropy will be at the maximum, and absolute disorder will reign. All visible matter — about half of the 5 percent of baryonic matter — is blowing away into disorder. What will finally remain are fields of energy and quantum flashes of positive and negative particles cancelling each other out.

## HUMAN ENTROPY

Let us apply entropy to ourselves. The chloroplasts of plants capture a tiny amount of the small portion of the sun's radiation that lands on our planet. The kinetic energy of the photons converts to chemical reactions to break down water, capture carbon, and combine it with hydrogen to make carbohydrates. By eating plants, or eating animals that eat plants, we consume and burn the results of the sun's radiation. By burning this energy, we maintain our structure and temperature.

We release energy locked within matter as we break down food. We use that energy in movement, and preserving our life, but most of the energy we release from matter is discharged in random particle movements as heat. Because our body is much hotter than our environment, we leak most of our energy as heat into the cooler environment. We contribute to the dissolution of order, which increases entropy, by a tiny, tiny part of an infinite process. Twenty percent of the chemical reactions that constitute our existence maintain our orderly structure and reproductive ability. The other 80 percent create entropy. That 80 percent encompasses our role in the cosmos.

All life forms throughout the universe will be found to contain chemical and physical processes to break down gradients and process complex structures, reducing them to their parts in random movement — just like us. Some portion of the process will be devoted to maintaining the life form. It will sense and interact with its environment, which also means processing information about its environment and itself and conveying some of the information to others of its kind. Its shape and particular chemical process, and the molecular structures it processes, will depend upon the environment. Given the universality of the entropic process, it is inevitable that life forms will be ubiquitous in the cosmos.

In conclusion, all forms of life, including our lives, are self-organized chemical processes that create entropy, seek the lowest energy state, and break down gradients to restore equilibrium to the universe. We are instruments to facilitate this process. In a world born to move from order to disorder, every step we take participates in a natural unwinding, a separation, with heat generating, scattering, and cooling, less and less interaction, and energy without work.

It is almost impossible for us to identify with the processes governing our existence because we cannot visualize the molecular and particle exchanges that constitute our world of motion and energy. Our limited point of view at our human scale causes our endemic confusion. The degree of magnitude between the human cell to the Planck unit (the smallest possible size of anything at 1.6 times 10 to the $-35$ meters across) is equal to the magnitude from the human cell to the end of the universe. We are unaware that we live in a storm of particles, and the storm of particles lives in us. Entropy is the scattering of the particles in this storm. The quantity and minuscule magnitude of the particles in the environment we live within is very difficult for our imagination to encompass; for example, with every inhalation, we breathe in nine sextillion molecules of nitrogen along with similar magnitudes of other minuscule molecules.

Meanwhile, we facilitate breaking down matter and scattering compact and organized energy into disordered, randomly dissipated heat by making a living, maintaining a shelter, engaging in social relationships, participating in our culture and institutions, sharing our beliefs, and having children. The habits, institutions, rituals, and above all the memes — intellectual concepts that organize our conceptions and perceptions of human interactions — enable our culture to form local environmental adaptations to fit our space and afford techniques and behaviors designed to facilitate entropy. Culture is Nature, as the mind is the body.

The greater efficiency of our technology and our emphasis on increased productivity, although it is "progress," can only increase entropy and waste. It is this contrast that befuddles our minds in this century as our ingenuity, technological progress, chemically enabled increase to food production, and population growth are confronted with increasing existential threats to the species in the running down of resources and potential for catastrophic changes. Our conundrum is: How when we're doing so well can we be doing so badly? It is said that humans destroy Nature. No! Humans *are* Nature. Humans cannot go against Nature, create less entropy, consume less, or be equal. We are machines to break down complexity.

We cannot escape our role in the unfolding cosmic drama as the universe seeks equilibrium by increasing entropy. The cosmos is in a non-equilibrium state because of the concentrations of matter/energy. The material cosmos is lumpy, comprising hot spots of concentrated matter within the cold vacuum of the void. The cosmic process is the smoothing of these gradients between high and low energy, eventually arriving at the lowest energy, which is the goal. All energy transfers from complex macroscopic objects of contained movement into the environment as random movement. It never flows back again into coherence. In the future, the energy/motion of the cosmos will accelerate the expansion of space, heat will approach absolute zero, and particle movement will virtually cease.

Ever-increasing entropy is our story at every moment, physically, economically, socially, and culturally. We define ourselves by the 20 percent of our energy that maintains our existence — albeit with great difficulty and temporarily. It is strange to us to think that on the surface we're spending our lives trying to create order in our lives whereas our actual role in nature is to do the opposite. We split ourselves in two conceptually with the concepts of good and evil: we label our search for meaning and coherence and order as good, whereas the breakdown of order and consequent randomness, we label evil. We go so far as to ascribe the origins of both as emanating from supernatural agents. Consequently, the entropy of waste energy and disorder that we live within and serve is removed from us, to become alien to us, and our enemy.

We are aware that we live subject to pain, uncertainty, and confusion while not knowing their source, but we suspect there is something hidden governing our lives. While swamped by pain and suffering we view the source as unknown and alien. Therapeutically, we minister to one individual to uncover the origins of the pain/disorder in their life, localized where it is appropriate in formative experiences, and gather information with which we can form an orderly narrative to focus our efforts. We attempt to replace delusion and denial with knowledge. Let us place entropy where it belongs as the central mechanism governing our lives as it governs the universe. The source of the major complaint that we do not realize the things we want most, although with all our heart we want to realize them, and instead do the things we do *not* want to do, although with all our heart we do not want to do them, is to be seen as an offshoot of our central identity as material beings subject to the laws of physics, and in particular to the second law of thermodynamics, which is the law of entropy. Therefore, we are faced with destructive and self-destructive behavior that is intrinsic to our nature.

We are inside, not outside, the laws of physics. Death is a transition of state with the release of entropy. This book describes how the human species exists within the second law. What can we do within these limits? Can morality coexist or even be derived from Nature's laws?

# Chapter 2

## From Beginning to End:
## The Cosmic Drama

A universe that begins in symmetry loses parity as lumps of matter form out of radiation creating potentials. However, matter and gradients only exist to be overcome and the force of entropy, and the return of matter to radiation restores the universe close to symmetry. The story that unfolds in this chapter is the story of humans. We are part of nature which is embedded in us as we are embedded in nature. We must follow the narratives of entropy and the cosmic process. We participate in the dissolution of coherent matter. Our narrative mirrors that of the universe's narrative and the universal law of entropy shapes every aspect of our lives socially, politically, economically, and culturally. We are totally within nature's story.

The universe is a process. Were it not for the second law of thermodynamics (entropy), any process would be fully reversible and whatever would come apart could be put together just as whatever was put together could come apart. The process in the arrangement of particles is primary and time is derived by our human perception, from relative vantage points, of the changes in the positions and velocities of composite matter. What process existed prior to this universe is unknown either mathematically or by experimental witness. However, a good guess would say there was no process, but there were intense fields of oscillating waves of energy infinitely squashed together at infinitely high temperatures. Without particles and composites there is no time. Why the fields separated is unknown.

We are constrained by our temporal category and therefore we tend to align things according to before and after rather than more fundamentally as a process with purpose. Therefore, any story we tell is before and after: this is followed by this, followed by this, to an end; there's always a beginning, even a beginning to the beginning. However, causality, including a first mover, is a concept that only applies to the universe after fields separated and particles existed as concentrations of the fields' energies. The quantum level describes the discrete movements of infinitely small particles within fields of wave-like energies out of which composites are processed, including us.

## THE PROCESS BEGINS

Before the event that created the universe — known as the Big Bang — there was a field of energy, uniform and in complete symmetry, comprising all that existed. There were no particles. Only after the symmetry was broken did the laws of physics begin, governing the relationships between the parts of the universe. Time began. The closest thing we have to "eternity" is the energy field from which the universe emerged with the Big Bang. There was no "before" before that.

Although science does not know the physics of the original energy field before it broke up into parts, the physics of the breakup is very secure because of the experiments examining particles, such as those at CERN, and because of mathematical proofs and reconstructions of the process that compare our present state with existing phenomena such as the cosmic microwave background (a remnant from the early stage of the universe).

The core of our difficulty in understanding the relationship between the quantum fields of energy that extend throughout the universe and the world of localized matter with defined position and momentum that we live in, is that our language evolved from our sensing apparatus, which is attuned to our ecological niche. Our language is ambiguous, our imagination limited, and the fundamental physical relationships defining our lives escape us. Only the language of numbers — mathematics — exactly defines the fundamental relationships. Such grave limitations are humbling for a species that thinks of itself as the center of the universe, exceptional in every way.

Nevertheless, an attempt can be made to translate the fundamental relationship between the fields that carry energy and matter into our language by using analogies, ending up with what physicist Niels Bohr said could only be poetry. The analogy I use is a comparison to the ocean's waves. A water wave transfers energy without transferring matter. The matter bobs up and down, initiated by the motion of the molecules within the wave, but the molecules themselves do not travel.

I think of the universe's initial state as infinitely compacted waves that have infinite energy and temperature because of their compaction. In the Big Bang, the waves spread out, increasing the wavelength and diminishing the energy intensity as their motion spreads. When it expands, it is revealed that there is not just one wave but several. These waves are force fields, which separate.

I think of the small amount of matter in the universe at the present time as eddies in the waves separated by their angular motion and spin. The eddies sometimes go in one direction and sometimes in another, cancelling each other out when they meet. They are also temporary, appearing and disappearing swiftly. In the beginning, not all the eddies disappeared, leaving behind local energy moving and spinning within a specific space.

What one has then are local movements (particles) that, although temporary, may last billions of years, but exist only within the non-local wave, which like the ocean is vastly extensive. The quantum force fields extend throughout space. Therefore, the particles exist within waves and act as both waves and particles. Matter is local, having a defined position and motion, but it is also located within a non-local energy, a wave that is coextensive with the universe. We are entirely grounded within the local and therefore do not directly experience the universe-wide waves that constitute the energy from which we derive our form. What we experience, like the eddy in a wave, is our position in local space and our extreme temporality. We appear and disappear (relative to the universe's age, now and continuing) as if in a fraction of a second.

The Big Bang created a universe with relativistic movements, such as photons and radiation, and also constrained movements caused notably by the strong force which contracts the energy of the fundamental particles (quarks) by the intense force-carrying particles called gluons and forms protons and neutrons at the core of atoms. Atoms in turn clumped to form composite, many-bodied matter. The matter coalesced because of gravitation, creating different levels of density and temperature. As we shall see, as the universe unfolds, eventually matter will be destroyed to return the universe's state close to its original symmetry.

## THE COSMIC STORY

The laws of physics define why life exists and how and why it evolves and shapes our lives. We are entirely governed by the vast and complex process of the entire cosmos. We live inside the cosmic process, and it lives inside us. It is the cosmic process that governs our lives, our relationships with others, our morality, and the reasons for our existence.

It is by recognizing our place within the cosmos that, therapeutically, we can revise the picture of our purpose and modify our goals individually and collectively. Instead of being hubristic, considering ourselves exceptional beings superior to all other beings in nature, we can realistically acknowledge our place and purpose. This opens the door to realistic strategies that will partially protect us from the impersonal forces that shape us. We need to adopt this therapeutic strategy to escape from our self-destructiveness, destructiveness, and delusion.

The cosmic process has stages: a beginning, a middle, and an end. The first stage took a few seconds. The storyline takes us from temperatures of quadrillions of degrees to the second stage, where our universe's current average temperature is a few degrees above absolute zero. Eventually, the cosmos will arrive at the lowest possible energy with a temperature barely varying from absolute zero.

## STAGE ONE: THE BEGINNING

13.7 billion years ago, at Planck Time, which is 10 million trillion trillion trillionths of a second, the energy of the original dot/wave/field vibrated at infinite speeds at a temperature of 100 million trillion trillion kelvins. Order was at its maximum, so entropy (randomness and disorder) was at its lowest. At 10 decillionths of a second, radiation spread out, creating space. There was no mass. As the universe expanded its density decreased, and the temperature marginally cooled. When the universe cooled the three combined quantum forces — the strong nuclear force, the weak nuclear force, and the electromagnetic force — separated.

The creation of space-time simultaneously created gravity, the fourth force of nature. Gravity is the ability of space and time to stretch and contract the relationship between objects. Gravity separated from the three quantum forces that control the relationships between the primary particles that constitute our universe. The strong force holds together the quarks, which are basic particles, in the nucleus of the atom. The electromagnetic force controls the relationship between electrically charged particles. The weak force allows exchanges in interactions between elementary particles through the decay of particles as in radiation, and the fusion of particles as in the fusion of hydrogen into helium in the thermonuclear process of the Sun. Three forces are quantum insofar as they describe relationships between the fundamental particles. Gravity is thought to be quantum, but it has not been mathematically or experimentally proven to be. Therefore, gravity is described through the relativistic theories of Einstein as a classical rather than quantum force, which means that it relates to the interrelationships of complex or many-bodied objects rather than fundamental particles.

First, gravity was separated from both the strong force and the electroweak force, a combination of the electromagnetic and weak forces. Then the strong force separated from the electroweak. The last separation of forces between the electromagnetic and weak force occurred after the universe inflated at a speed greater than the speed of light, repeatedly doubling in infinitesimal time. The electromagnetic force carried by the photon then separated from the weak force.

With further cooling, at about one-trillionth of a second after the Big Bang, the quark-gluon plasma that emerged from excitations in the quantum fields began to break up. Quarks are the four elementary parts of matter and make up protons and neutrons, which in turn will become the core of atoms when combined with electrons carried by the electromagnetic force. A particle appropriately called a gluon carries the strong force, which confines the quarks to create matter. The combined quarks formed both matter and antimatter, quickly annihilating each other to produce pure energy. The annihilation should have left a universe with only radiation and no matter — or us. However, one particle in a billion survived. Parity was broken, creating the slight residue of one in a billion that populates our universe with the baryonic matter that constitutes all the stars, planets — and our bodies.

Next, the energy density of space created a pressure that inflated space faster than the speed of light. In a billionth of a billionth of a billionth of a billionth of a second, the cosmos (all of space) expanded from little to somewhere between the size of a grapefruit and a small galaxy. The energy of the inflation field then decayed into electromagnetic radiation. The universe smoothed out, except for quantum fluctuations, and was flat, symmetrical, and almost empty as the density of particles and radiation decreased because of the expansion. After inflation, the electromagnetic field separated from the weak force, and the Higgs field gave mass to some elementary particles. The Higgs field interacts with some particles to resist their movement in speed and position. It is this resistance, which can be compared to trying to move through molasses or trudging through mud, that causes the feeling of mass. The field is mediated through a particle called the Higgs boson. Without the Higgs field there would be no mass in the universe.

In the initial singularity, the relativistic movement of photons was constrained. The cosmic expansion by inflation forces stretched the frequency and wavelengths of radiation thereby decreasing its intensity. The spreading-out of radiation dramatically decreased the temperature of the universe.

Between one second and ten seconds, the leptons — negatively charged electrons and positively charged positrons — were in pairs, but with a decrease in temperature, pairs were no longer created. The initial leptons mostly annihilated each other, giving rise to high-energy photons. Once again, parity was broken when a slight residue of leptons, primarily electrons, was left.

The intensity of the radiation at first prevented quarks from grouping under the strong force to form the protons and neutrons at the core of composite matter. However, when the expansion of space diluted radiation, quarks combined under the strong force to produce protons and neutrons. The early universe was then a gas, a plasma of thermal radiation, electrons, protons, neutrons, neutrinos, and dark matter.

The explosion, and probably the inflation, was caused by the outward pressure that exists in everything that is compacted. Gradually gravity together with the aid of dark matter began to coalesce the particles of the explosion into the stars and galaxies. Gravity slowed the universe's expansion. However, once the density stretched out, gravity became weaker because of the increased distance between the stars and galaxies. At that point the inherent pressure of the contained body, which is the universe, took over, and continues to accelerate and accelerate the distance between objects as gravity weakens.

## THE MIDDLE STAGE OF THE PROCESS

The middle of the story begins with the slow transformation from gas plasma to the dominance of matter. Three minutes after the Big Bang, the universe was 333 million times more compact than today, with seventy times the current temperature of our Sun's core. The components of matter were 75 percent single-proton hydrogen nuclei, 25 percent helium nuclei, and small amounts of lithium-7.

After 47,000 years, because of the expansion and cooling of the universe, the density of matter and radiation were equal. By 370,000 years, the temperature was 4,000 kelvins. At that temperature, electrons could bond to the atomic nuclei of protons and neutrons to form the neutrally charged atoms that, being at the core of matter, are the constituents of us.

For the first million years, radiation and matter were in equilibrium. Then, matter began to dominate as free-streaming radiation spread into the expanding space and density was diluted.

At first, the universe was expanding too rapidly for structures to form. As it slowed, tiny pockets of greater density created gradients in a predominantly uniform gas and became seeds for larger structures. Matter comes in two varieties: baryonic and dark. Baryonic matter is made of protons and neutrons, which interact with the electromagnetic force field, emitting photons that make it visible.

Dark matter does not interact with the electromagnetic force. It shows its existence and is only quantified as it interacts with gravity. Radiation pressure, created in interaction with the electromagnetic field, slowed the concentration of baryonic matter. However, dark matter is not slowed down because it does not interact with the electromagnetic field. Therefore, it collapses faster. Gravitational instability drew dark matter together, so the tiny irregularities in the universe's density were amplified, creating structures of dark matter filaments.

Although dark matter forms structures, unlike baryonic matter it cannot clump because it does not lose energy through radiation. Instead, it forms vast, diffuse filaments. By contrast, ordinary baryonic matter loses energy by radiation, causing it to form dense objects and gas clouds.

Because dark matter is quantitatively greater than visible matter, it was the primary force in the formation of galaxies. Between 380 and 700 million years, hydrogen and helium atoms clumped together to form stars and galaxies as they fell into dark matter through gravitation. Dark matter's gravitational pull continues to prevent galaxies from flying apart in their spiral rotations.

When atoms are jammed together, their movement intensifies, creating heat. Under pressure and heat, hydrogen ions (an ion is a nucleus without electrons) fused to make helium ions. When that happened, even a small amount of mass when converted to energy and multiplied by the square of the speed of light, blasted out as electromagnetic radiation to light the stars. Before that time, the sky was dark.

Stars continue to convert matter to energy during fusion and scatter radiation into the void.

A star's density increases as it burns. The increased density, in turn, magnifies the heat and creates more nuclear fusion. The greater density, heat, and fusion create the heavier elements of carbon, nitrogen, and oxygen. Eventually, a star's density and heat initiate an explosion spewing condensed heavy elements throughout the universe. These heavier elements form complex structures such as the planets and us.

The universe is at least 91 billion light-years across. (Ninety-one billion light-years is the distance a photon will travel over 91 billion years at a speed of 186,282 miles per second.) Between 100 million and 200 million galaxies contain, on average, about 100 million stars each. Some galaxies have 100 trillion stars. Ordinary matter is about 4.6 percent of the mass of the cosmos, but half of the baryons are scattered as atoms throughout the universe. The rest forms all the matter that we can see. By comparison, dark matter, which we only know from its gravitational effect on galaxies, is 26.5 percent of the universe's mass.

## THE END OF THE PROCESS, NOW IN PROGRESS

The end of the story began 9.8 billion years ago when the universe, which was gradually decelerating because of the effects of gravitational attraction on the baryonic matter, surprisingly began to accelerate. The large conflict within the universe is between gravity and dark energy. As the galaxies recede from each other, gravity becomes a relatively less powerful restraint on matter compared to the pressure of expansive energy. This energy, termed dark energy, which dilutes matter's density, makes up 68.3 percent of the universe. As baryonic matter burns up, matter will completely cease to exist after a vast period and the energy from the converted matter will process the expansion of the universe.

We have only existed for the last few million years and are characters in an extended final chapter. We will cease to exist before the cosmos completes its arc. Our demise happens soon in cosmic time. In 600 million years, the Sun's luminosity will have increased to the point where carbon dioxide decreases, plants will no longer be viable, and therefore animals will die. (Our lasting 600 million years is most unlikely either because of our self-destructiveness or because the coming together of the continents will create a hostile environment.) In four billion years, long after our species' death, the Earth's surface will melt, and the last remaining life form, bacteria, will be killed.

As space expands between the galaxies, they will stand alone. Here is what happens. The stars will burn out. A typical star like our Sun fuses hydrogen atoms into helium and converts 4.26 million tons of matter into radiation every second. The radiation, pressing from the core to the surface, enlarges the Sun. The burning of the hydrogen leaves a helium core. As the helium burns and the temperature increases, the core is reduced to oxygen and carbon. The outer layers of radiation expand, creating a bloated red giant. The outer layers gradually dissipate, leaving a white dwarf, a solid object that slowly disintegrates. If pairs of white dwarfs collide and merge, or the core of a giant star collapses, a supernova explosion occurs. Even more massive stars can morph directly into neutron stars or black holes. A star explodes every second or so in the universe. Even the ultra-dense black holes give off radiation and eventually wholly evaporate.

The universe is between 13.77 billion and 13.82 billion years old. The end of composite matter may come in 5 billion years. All baryonic matter will become an infinitesimally small portion of the cosmos. Space will continue to expand at increasing speeds. Individual atoms will be ripped apart at 22 billion years.

We live in that portion of the universe's life during which composite matter exists. However, the fate of the universe is that matter is burning up and being converted into radiation as disordered energy. The cosmos will then have no ordered structures, complex processes, or internal movement apart from quantum virtual particles. The temperature will be barely above absolute zero.

Everything within the universe is being converted to the lowest energy state and random movement, while all energy serves the universe's expansion. This purpose is independent of the time it takes. Ten years or ten billion years are meaningfully the same. Process is purpose. This movement is our process and our purpose.

The process of the universe is to restore thermodynamic equilibrium and to rest at the lowest energy state. The potential energy in complex matter must be restored to radiation and will function exclusively to expand space.

Gradients will be eliminated. The law of entropy (the dissolution of all matter) conquers all variations in density and all temperature gradients.

## OUR PURPOSE

Life — all of it, including us — is a set of chemical processes that evolved to facilitate the breakdown of gradients. We were assembled as part of this process, making it our purpose. Our human story, individually, socially, and institutionally, mirrors the process/story of the universe. We metabolize — burn up — complex molecules and exude electromagnetic radiation. We break down ordered complex matter into random particles of heat/energy. Our contribution is negligible, but what applies to the cosmos applies to its contents. Aristotle said the purpose of a thing was to do what it was best at. We are best at breaking down ordered matter into disordered movements, creating entropy. We are 80 percent efficient at our purpose. Humans are embedded in the cosmic process.

As the book progresses, we will see that the process of the cosmos, the arc by which it comes into being and then leaves, is integral to every aspect of our lives, including our culture. We are contained within it, and we are made of it and to its purpose. If the species were an individual, we could compare it to someone who is born into adverse circumstances, such as someone who is a Dalit in India treated as an untouchable of the lowest class, or Black in America, or someone who has genetic difficulties that limit their functioning. Therapeutically fostering a delusion that all is well or will be well would be irresponsible. What is required is to evaluate our limitations and to create a strategy that ameliorates our difficulties. We already try to do this, but still we subject ourselves to the destructive and self-destructive behaviors that are entailed by the physics of the universe, and counter them ineffectually by favoring ineffectual strategies or by composing delusionary stories that turn us away from effectually addressing our quandary.

The entire gravitational pull of our planet does not prevent us from moving with alacrity when we are young, trudging a bit when we are older. Similarly, the cosmic drama does not prevent us from moving freely when we are young, with difficulty when we are older. Increasingly we are aware of how intense the struggle for life is, which is cloaked in danger. Ultimately, we face our extinction. Our personal journey mirrors that of the universe.

Therapeutically, by recognizing our place within the cosmos, we can revise our beliefs in our purpose and modify our goals individually and collectively. Instead of considering ourselves superior to all other beings, we can realistically acknowledge our place and purpose in the cosmos. This opens the door to realistic strategies that could partially protect us from the impersonal forces that shape us. This must be part of the therapeutic strategy, which the species needs to adopt to escape from its trap of destructiveness, self-destructiveness, and delusion.

# Chapter 3

## We Are Made of Bits and Pieces

We are named at birth, given an identity that affirms we are whole. We see ourselves and we are seen by others grossly. But what is it to be solid, to be whole?

It is only in the past few centuries, and above all in this past century, that we have generally become aware of our microscopic parts as we get blood tests and talk of cancer cells, brain cells, and gut bacteria. We talk of how these parts work together, interact, and make us whole.

At the subatomic level, our wholeness is reliant on forces and fields. One force holds infinitely tiny bits and pieces of us together, while another field gives them mass, and another attaches the bits. Another force compacts them into even larger bits, while the weak force helps them to interact and fall apart.

By thinking of ourselves as a whole, we place ourselves apart from our constituents, from nature and its laws. We claim a place above and beyond everything else in the universe. However, the interactions between our microscopic parts ground us firmly within nature and control our functioning and behavior. We experience ourselves as solids and participate in a world we think of as solid. However, we solely depend upon the primary forces, the fundamental realities that constitute the universe.

From our macro point of view, we do not know how we function, but we sense that we are falling apart. From early times, as evidenced in the horror of death in the Sumerian epic of Gilgamesh, we have been haunted by the contrast between the conviction of our power and the precariousness of our existence. Two thousand five hundred years ago, Buddha said: Everything is made of parts and, therefore, will fall apart. Don't cling. It is clinging that causes pain.

## HOLDING TOGETHER

The body has 1,000 trillion trillion atoms. To get an idea of the scale of the particles within an atom, it has been suggested that we picture a fly within a cathedral. The cathedral represents the atom, the fly the nucleus within an atom. Positively charged protons or neutrons without charge are nuclei. They are surrounded by electrons, which exist as a cloud of negative charge in an orbital energy shell. The electron's location is only statistically apparent. An electron is about one ten-millionth of an atom. A proton is one trillionth of the atom's volume. Its mass is 1,836 times that of the electron. The rest of the atom is a force field empty of matter. Within the proton and neutron, the most fundamental particles are the quarks, which are mediated by the strong force carrying gluon, holding them together. If you took away all the empty space between these elementary particles, it has been calculated that the elementary particles of the entire human race could be contained within the size of a sugar cube.

Even more fundamental are fields and forces such as electromagnetism and the strong force that hold the quarks within the nucleus. The proton's mass is the contained kinetic energy (movement) of the quarks. Particles are excitations of fields. The lowest wave frequency in a field is the minimum energy, and its energy corresponds to its mass.

Electrons are exchanged and shared between atoms, creating bonds and making molecules. Complex molecules are made when they stick to each other by sharing electrons. This process creates chemical reactions in and between objects. It is this sticking together of atoms and molecules by sharing electrons that underpins our existence.

## FALLING APART

We not only live in a storm of particles — trillions upon trillions upon trillions — we *are* a storm of particles. Our cellular structure is composed of tiny particles made of tinier particles. To acquire our energy/movement we must constantly harvest electrons by breaking down matter. Thus, we create entropy by eliminating the gradient between concentrations of potential bound energy and unbounded kinetic energy. We are constantly buffeted by a storm of particles within us and around us. We are fragile and desperate. Eventually we cannot keep up as the processes within us, and the parts, fail. Movement slows and we fall apart.

The primary element in living things is the cell, and cells communicate with each other to form multicellular structures with singular purposes. Multicellular structures form organs, and then organs link up with others to form more complex living beings. Humans are a collection of cells which are differentiated according to different functions but work together to break down complex matter into the random particle movements of heat. We are solids that comprise communities made from trillions of cells living cooperatively, and trillions of other cells such as bacteria which also inhabit the community. We also live in a sea of trillions upon trillions of molecules, atoms, quarks, forces, and fields both within and around us.

Life is a system of chemical reactions that dissolve minerals, rocks, simple cells, and multicell creatures while exuding methane, water, oxygen, and carbon dioxide. These reactions break down complexity and vent heat in all life forms, including humans.

Collections of particles push against other complex amalgamations to perform work. "Working energy" measures the force and the changes that pushing makes. Work enables particles to form more complex combinations and objects. Work in the cell breaks down other cells and molecules to harvest their contents.

Within our membrane, we are 60 percent water and 87 percent gases, namely oxygen, hydrogen, and nitrogen. Most of the hydrogen and oxygen are in the water. We carry around the ocean from which we emerged, and our lifeline is oxygen. Nevertheless, we do not think of ourselves as primarily liquids or gas. Thanks to the fields and forces, we experience ourselves as solids.

All things exist in a scramble. The scramble is the constant alignment and realignment that seeks stability. All things are constantly faced with fragmentation. Just look at atoms seeking to retain balance between positive and negative charges, filling the outer shell with electrons and bonding with other atoms to create molecules. Everything is in motion, and there is motion in everything. Energy is everywhere, and mass and energy are interchangeable.

## ON THE BRINK

Our macroscopic self, when looked at closely, is an illusion. We and all other solids are contingent — a skim on reality. We say, "He is solid. You can trust him." "She has it together. She won't fall apart." "He was strong, and now he's becoming so frail, just skin and bone." We are solid, but we worry. We are stable for a short time but always coming apart, falling apart, failing, declining, losing it, losing it all. We are in denial, claiming we are okay, assuring ourselves. But we know.

We are mostly space, and we fall into it. Our bodies are in flux at the microscopic level. There is a chasm between the subjective and the objective pictures of ourselves. We are quantum creatures. We exist with vertigo on the edge of our nothingness. We process in death to no weight at all. Our fear is not just of our death but of our extinction. We dance on emptiness, like bubbles on a stream.

Our seeming solidity is grounded on our means of perception through photons. Fields, electromagnetic waves of various lengths, neutrinos, and gravitational waves travel through our bodies as if we are not here. The distance from a cell to a quark is equal in scale to that between a cell and the edge of the universe. Looked at one way we exist while looked at in another way we do not. Our boundaries are permeable.

I symbolically think of us as equilateral (perhaps because of the three quarks). We usually think of a triangle resting securely on one of its sides. However, I envision it resting on its angle, becoming tinier and tinier as it rests on its foundation — a point. We are wobbly and always on the tipping point.

We are like virtual particles, ghostlike, winking in and out of existence. An individual's lifetime in relation to the universe's life is the equivalent of the nanoseconds of a virtual particle. One can think of human lives as quantum existences with statistical probabilities based on genetic irregularities and randomness. The random moment an individual exists in depends upon their place and time of birth.

Humans believe the world and indeed the universe revolves around them and was created for their benefit. We think of ourselves as apart from nature and supreme. However, there are other traditions. In the *Iliad*, humans are the playthings of gods. In the Bhagavad Gita, all things are an illusion. Ecclesiastes summarizes human life as "from the dust we came and to the dust we shall return."

Similarly, in Shakespeare's *The Tempest*:

Our revels now are ended. These our actors,
As I foretold you, were all spirits, and
Are melted into air, into thin air:
And like the baseless fabric of this vision,
The cloud-capp'd tow'rs, the gorgeous palaces,
The solemn temples, the great globe itself,
Yea, all which it inherit, shall dissolve,
And, like this insubstantial pageant faded,
Leave not a rack behind. We are such stuff
As dreams are made on; and our little life
Is rounded with a sleep.

Physics and its chemical expressions belie our Myth of Human Exceptionalism with the reality of our entropic insatiability. The drive to remove gradients is written into our cellular structure and expresses itself in our insatiable compulsion to consume as much as possible. The inescapable factors are our transitory nature, ill-defined identity, desperation for electrons, and inherent fragility.

Why not embrace our insubstantiality? We are not God's image, the universe's consummation, or the measure of all things. We live within the quantum laws of nature as a chemical process reducing matter to radiation. Our effect on the unfolding process of the cosmos is vanishingly tiny. The unrelenting push to emphasize our greatness separates us from nature and an awareness of the fragility of our constitution.

Our species is equivalent to a neurotic individual in a state of denial pretending to be an exception to the natural world, superior, capable of defying entropy. Living in denial is at the heart of our claim to substantiality. The consequences of living in denial are that we must constantly fake it and are vulnerable to beliefs, political and religious, that are fake. The more they are fake the more we must insist on their reality. We unleash intolerance and violence.

The therapeutic is the recognition of our actual state. As education shifts toward physics, mathematics, and data analysis, we gain humility by thoroughly understanding our constituent parts and their functioning. On the other hand, this shift toward a secular view of life is seen as a threat to our identity and a dangerous deflation to be fiercely resisted. As in therapy with an individual, resisting reality summons a panoply of defenses.

# Chapter 4

## We Are the Product of
## Other Species' Parts

What a thing is made to do is its purpose, and its purpose is its meaning. On the micro level, our origin and meaning are present in the first living cell, which was chemically assembled to break down the density of matter. Cells then evolved to burn up matter in diverse environments and release the energy as random molecules and radiation. We are a collection of cells — matter continuing to serve the original purpose for which cells were assembled.

On the macro level, when we think of our species origins, Charles Darwin and the theory of evolution comes to mind — that is, the way that species have changed over time, adapting to their environments, and preserving their species by exploiting the random changes to their DNA. However, on the micro level evolution is better described as the creation of molecular efficiency in the processing of matter into energy. The determining factor on the micro level is the creation of entropic efficiency, the elimination of gradients, and finding the lowest energy level. The driving process in evolution is physics.

## LIFE ON EARTH BEGINS

We are mind-bogglingly recent. The Earth came into existence 4.57 billion years ago, 9 billion years after the creation of the universe. The first single-celled living creatures — archaea and bacteria — emerged in water between 400 million and 800 million years after the Earth's formation.

How did life happen? Scientists think that life on Earth may have begun about 4 billion years ago with ribonucleic acid (RNA), a molecule present in most living things. Ammonium sulfide from volcanoes in the early Earth contained the nitrogen that formed the core of the four nucleobases of RNA: cytosine, guanine, adenine, and uracil. These nucleobases (nitrogen-containing compounds) formed a chain by the covalent bonds of shared electrons between the sugar deoxyribose of one nucleotide and the phosphate group of the next. The RNA sequences contain the steps for forming proteins, which catalyze the construction of cells out of amino acids. Then RNA evolved the ability to replicate.

There are two theories about where the self-organized strand of molecules that could copy itself first happened: either under or above the water. At this time, Earth was a "water world," covered largely by ocean. Magma from the Earth's core seeped through the seabed, creating large, mineral-rich underwater structures. Concurrently, volcanoes peeked their heads above the oceans as they do in Iceland today.

The first hypothesis is that hydrothermal pools on volcanic islands underwent wetting and drying cycles that concentrated nitrogen, phosphorus, and sugar solutions to form strings of repeating nucleic acids. According to the more plausible second hypothesis, deep-sea hydrothermal vents, which are hot underwater structures rich in sulfur, iron, and mineral particles, catalyzed the creation of simple organic molecules with cell membranes.

## LIFE DEVELOPS

Life and our experience of it is the movement of electrons between atoms and between molecules. To create life by breaking down more complex objects requires this movement. An atom or molecule donates electrons because it has a sufficient number, while another accepts electrons because it is deficient. Initially, archaea obtained their energy from inorganic compounds by moving electrons between electron donors such as hydrogen gases, sulfur, ferrous iron, and ammonium, and electron acceptors such as carbon dioxide, sulfate, nitrate, and ferric iron.

We are the sum of chemical processes creating equilibrium between gradients — differences between hot and cool, differences in the concentration of atoms, and instabilities within molecules and atoms that seek a lower stable energy state. Life's physical functions arose from random and evolving chemical processes.

The cell is the basis for all living entities. Cellular species are replicating chemical creatures that initially transformed matter to energy by stripping out electrons by using sulfur as the electron acceptor that catalyzed the process. Cells are the tiny animals that do nature's business of breaking down density and the concentration of energy in matter. Essentially, we are collections of these tiny machines.

Unlike human cells, archaea and bacteria lack a nucleus containing DNA. The road to establishing cells with a nucleus began between 2.3 billion and 2.7 billion years ago, when cyanobacteria (blue-green algae) began taking carbon dioxide from the air and using the energy from photons (sunlight) to separate the carbon from the oxygen. The algae structure needed carbon, but oxygen was poison to its nitrogen-fixing enzymes. Oxygen was vacated into the predominantly methane atmosphere. This process of using sunlight and carbon dioxide for energy and releasing oxygen as a by-product is called photosynthesis, and resulted in an atmosphere that became richer in oxygen over the next 400 million years. The abundance of oxygen provided the foundation for the next stage in the evolution of cells.

The next significant step in the creation of cells that make up the animals and plants present today took place about 2.2 billion years ago, when a single-celled prokaryote (Asgard archaea) swallowed a bacterium (mitochondria) to form a symbiosis. The archaea provided the mitochondria with the chemical molecules they needed, while the mito-chondria processed the archaea's food into energy through chemical reactions that exchanged electrons using the newly abundant oxygen.

This new eukaryote cell is the foundation for almost all living things because oxygen's need for electrons is a superior energy-extracting capability. The increased energy enabled a radical evolutionary leap from unicellular to multicellular creatures. Cells eventually communicated and cooperated to form communities that evolved into various life forms, including our human cellular community.

The cell's parts included a membrane, chemical exchanges (feeding), and often an appendage that fluttered to create locomotion. Some microscopic single-celled creatures formed mat colonies covering the ocean floor. They messaged each other through molecules. Single-cell microbes, the largest of which were grain-sized, are our building blocks.

Algae continued to expel oxygen into the atmosphere over hundreds of millions of years until, only one billion years ago, there was sufficient oxygen to maintain larger bodies. The movement of electrons driven by electron-hungry oxygen enabled the construction within cells of additional proteins and lipids, creating new moving parts that digested food, and other parts that communicated. Some of these multicellular structures were rooted in place as fronds and ribbons, and others crawled. It took 400 million more years before sponges and jellyfish moved freely.

Five hundred million years ago, modifications in the anatomy of worms and fish signaled complex anatomy. The symmetrical plan of a head, four limbs, a digestive tract, and a nervous system organized from front to back and side to side is written in our genes from these early times. Our hands and legs are modified fish fins. Small worms without heads had nerve cords — the beginnings of a nervous system — and the cartilage that supported fish gills morphed into bones.

At first, waste was regurgitated through the mouth, greatly restricting the amount consumed and the energy available. The development of the anus opening permitted more food to process through the body, increasing body size because of the greater energy input. When fish went onto the land and evolved into reptiles 360 million years ago, a flexible neck allowed the head to rotate without moving the entire body. Sounds created vibrations in the jawbone, which helped form the inner ear. More finely tuned hearing facilitated the identification of food and predators.

As a result of progressive developments, 190 million years ago — relatively recent times — tiny shrew-like creatures became the first mammals to feed their young through glands. Our body, organ plan, and functions are like our little rodent ancestors. We are an assembly of parts. We have become accustomed to thinking of ourselves as primates, but we are distant cousins of rodents. It would help us if we imagined ourselves as very clever and much larger rodents.

## OUR CELLULAR SUPERCOMMUNITY

We speak of humans having cells, but from an equally viable perspective, we can say that three trillion cells have a human. Cooperatives of cells encompass all the functions we call a human. We are a supercommunity of three trillion complex cells attached, communicating, and performing specialized tasks. Organ assemblages communicate by nerves and hormones with other living cooperatives within our skin envelope to form a larger solid. We are a community of communities.

Our senses register some of our more significant structures. However, we have almost no awareness of the liver, kidneys, arteries, lymph nodes, spleen, and individual parts of the nervous system, let alone the cell functions, where mitochondria — the ancient bacteria — supply energy to our body. We merely feel energy or weakness.

The human body's confederacy of communities constantly changes and interacts with the environment until it falls apart. From another perspective, a human is an information machine with a set of instructions — an algorithm.

"Self" is a name encapsulating our ignorance. The descriptor "self" exists because of our limited awareness, which is focused on predation, mating, and avoiding being eaten. Our sense of self is a consequence of the grossness of our awareness. We do not have to know our functioning or how our body is seamlessly connected to nature. We are fully integrated into the physics and chemistry of our environment, internal and external, but can articulate only a small part from a limited point of view.

Our culture functions for us as the culture of a crow does for a crow, or an elephant's, a dolphin's, or a squid's. We are different in the chance alterations to DNA that allowed each of our species to harvest energy in different ways and in different niche environments.

We do not consider our body and its energetic functioning as our fundamental reality. Instead of seeing ourselves as evolved in and by nature as solutions to the processing of matter and creating entropy, we see ourselves as magically created like Athena was sprung from the head of Zeus. We flatter ourselves. Although we now know how composite we are and how we are related to all other cellular creatures, we persist in acting as if our bodies and the physical world exist only to carry our ambitions, beliefs, language, science, explorations, religious speculations, minds, and souls. We treat the body as a donkey that carries its master.

Let us turn this upside down. What is meaningful is our immersion in the process of nature that we share with the other species as we break down gradients. Our body plan and functions compare closely to other species'. We have inherited the environmental adaptations of earlier species. We are embedded in physical and chemical laws. Culture, language, and stories beguile us because they are our specialties. However, they are local adaptations, not what drives our evolution and functioning.

All our parts have been bequeathed to us by prior species. Our continuity with all other creatures is evident in our DNA. We share 90 percent of our DNA with mice, including 97.5 percent of our working DNA even though we separated from them seventy million years ago. Remarkably, we share 60 percent of our DNA with fruit flies. Their anatomy and circumstances are unlike ours, but modified and repurposed; we have re-adapted their genetic solutions to work for us in our niche.

Environmental changes precipitate life-form changes that preserve the accidental genetic changes that increase efficiencies in resolving gradients and metabolizing matter in their altered environment. That is all there is to it. Life is a small chemical creation making better living through chemistry in a tiny local world.

We carry our planet's history from the developing inorganic and organic compounds to evolved complexity, from one cellular being to the next: bacteria to amoeba, slug to fish, shrew to human. Our genetic assemblage was put together similarly to the machinery we manufacture in different places and assemble in separate factories, each adding small bits with various controls, functions, and labels.

There is a troubling contrast between the almost infinitely slow developments in evolution to the dominance of our species in the brief period of 300,000 years. Most of our triumph has come in the last 12,000 years with the domestication of plants and animals, the last 6,000 years with the concentration of populations in large towns and then cities, and the dramatic, exponential increase in our understanding of the world over the last 200 years because of science. During the last 70 years, our species has changed the planet's fundamentals by the dramatic increase in population and the exploitation of fossil fuels that has accompanied the inventions resulting from science. We initiated the new epoch of the Anthropocene. Our single species has hijacked the planet.

We came from inner rather than outer space, from the parlaying of random changes in the structure of our brain into world dominance because of new ways of imaging and communicating. Humans are a twig on a branch of the

Tree of Life. Will the bough break? Is the twig infected? If so, will the infection spread to all the Tree of Life?

What is the disease? On the one hand, there is no disease, just the drive to break down matter and smooth out the concentrations in density and temperature. It is an unreflective process which will be fulfilled when the Earth and our species are destroyed.

On the other hand, we cannot be satisfied as a species with this answer even though it is true because locally, we are embedded in the culture that we have created in response to our environment. We strive to function as individuals and families and maintain a modicum of safety prior to our falling apart.

Thinking locally from within our narrow human viewpoint, then, what is the disease? We believe we have been given privileges and special treatment as mirrors of divinity, even if we are debased. We glory over and have contempt for others in our community of living species. We have the right to exploit them to get whatever we want. Even within the groupings within our species, we belittle other groups to make ourselves appear superior. Whatever group we belong to, as well as in respect to other species, we believe that we are unique. We tend to see ourselves as apart from others rather than cooperative, both among ourselves and with respect to other species. We are cruel, lacking the ability to identify with others, and absent empathy. Our species is emotionally mercurial, easily angered, and constantly stressed. The species may be suicidal. As a species, we are preoccupied with our fantasies about our power, our brilliance, and how wonderful, successful, and singular we are. To characterize it psychologically, we could say that our species is highly narcissistic. Our systemic beliefs are pathological since we are self-destructive.

Our species' narcissism is a defensive reaction to our self-doubt, fear, and failure, which is caused by the law of entropy, the direction of the cosmos, and to the marginal role we play within it. To compensate, we grossly inflate ourselves.

## CAN WE CHANGE?

As individuals we find ourselves repeating behavior that causes us pain without understanding why we are not in control of ourselves as the captains of our souls. Paul (paraphrased) complained in the book of Romans, "That which I want to do with all my heart I do not do and that which I do not want to do with all my heart that I do all the time." Resolving this conundrum is at the heart of the practice of psychotherapy.

The same complaint can be made by us as a species. We exist within tragic repetitions without understanding. As individuals we are largely oblivious of the formative dynamics within our families as well as within our local communities and cultures, which restrict our beliefs and behaviors. Becoming aware of these formative influences may alter to a certain degree the balance within the personality of the individual as different beliefs are weighed differently, some becoming more credible and influential and some becoming less. Even small changes that we make in the direction of our life, the environments we choose to live in, or the values we inherit or choose will lead us in different directions than our upbringings that conditioned us. The results become amplified over time as we diverge more and more from our roots and initial direction. Moreover, if we are flexible, as we move down different paths, we find other paths that are even more promising and so it is possible by incremental steps to end up in very different places than the ones we first envisioned. This commonplace should give us hope that small changes might result in significant results more conducive to our safety. This is certainly the experience of the reflective process of psychotherapy — the examination of the sources of our beliefs, testing them against reality.

The same may be true in a therapy of our species. The species' direction can be tweaked by acknowledging the scientifically proven sources of our behaviors and the realities of our purpose and meaning. In this chapter, focusing on evolution, we find that we have purposes we have not considered and have been formed by influences which ground us firmly in the developments of our family of cells, past and present, and our rodent as well as primate ancestors. Our behaviors have been conditioned by the purposes and behaviors that have been instilled in us from the original development of life in the cell, but these purposes do not diminish our pain or grant us safety during our struggle for survival in a world governed by entropy. Alternatively, we may continue to think of ourselves as aliens from outer space, chips off the eternal power outside the physical, mind rather than matter-based, God-connected, born from an eternal world to which we will return.

We are at a point in our development when we have two models to explain our nature, one of which is scientific and deeply embedded within our structure and the cosmos, and the other is an ancient part of our storytelling culture by which we structured our societies to survive within our environment. Underneath our species' narcissism is fear, self-doubt, and inevitable failure caused by the law of entropy, the direction of the cosmos, and, to whatever degree we intuit it, our marginal role.

In this chapter, I have proposed that we are not isolated but in fact are integrated with the development of other species, as has been demonstrated in the science of evolution. Our integration extends to the earliest life forms and exists at the molecular level both in our development from the past and in our constituents in the present. To save ourselves, we need to abandon our sense of superiority and separateness. If we can integrate with our biological unconscious, despite the mandate to create entropy at our core, then we can find safety for ourselves and bring safety to others, at least in the short term.

# Chapter 5

## The Life Force Is an Electron: We Are Chemical Creatures

What is spirit? Is life a non-physical presence, a gift of God, or a spark? Scientifically life is chemistry, the combination and separation of molecules, until it is physics in the form of released electrons. The gradual release of four high-energy electrons carried by the molecule ATP — constituted and reconstituted trillions of times over within the mitochondrion membrane — is energy and life. We live in a world of electrons.

Our human energy, life, and spirit are created solely from the breakdown of composite matter with most of it vacated into random movements as heat. Twenty percent of the energy released by us in the process is reserved to continue our existence (for a short time), while 80 percent is vacated. This is our destiny.

# LIFE'S PROCESS

A covalent bond is created when a pair of electrons are shared between two atoms. The atoms are bonded by the electrostatic charge of their nuclei to the same electrons. An ionic bond occurs when one atom's electrons migrate to the partially empty orbit of another atom. The atoms become bonded to each other because of the opposite charges. When either of these processes happens, the atoms become oppositely charged and bond. The result is a neutrally charged molecule. In the same way, two or more molecules can bond together. When the bonds break, the electron energy that the bond constrains is released. Each energy release is tiny, but there are trillions of changes constantly happening. The daily movement of trillions of electrons in our bodies is our energy, "life force," spirit, and aliveness. The bonding, breaking of bonds, and electron flow occur in the vast storm of particles swirling about inside and outside our bodies, buffeting the processes that struggle to maintain order and create work.

ATP is the molecule that carries energy to every cell and process. Each human cell needs 10 million molecules of ATP a second. Multiply this by 37 trillion cells in the human body! Once or twice a minute, each ATP molecule is recharged. To understand our subjective experience of being alive, we must closely examine how ATP is generated and regenerated.

Plant photosynthesis produces glucose, stored as a cellulose polymer in cell walls. Animals break cellulose down by chewing and digesting the plant's cellulose to obtain glucose, which is made usable by the metabolic pathway of glycolysis. Glucose is combined with a phosphorus group to prevent the reaction from becoming reversible. Phosphorus plays a central part in the entire process of generating energy. Enzymes chemically catalyze the rearrangement and bonding of molecules to produce citrate and the key oxidizing enzyme NAD+ (nicotinamide adenine dinucleotide), which accepts and carries electrons from place to place.

The next step is the citric acid cycle, which occurs inside the mitochondrion organelle. Electrons from the citric acid formed by the previous step are transferred to NAD+ (the acceptor of electrons), transforming it into NADH. Every acetyl group of the citrate produces three molecules of NADH. NADH carries the electrons into the oxidative phosphorylation process. This process produces 7.5 ATP molecules. An additional coenzyme generates another 1.5 ATP.

Four protein complexes within the inner membrane of the mitochondrion perform the harvesting of the electron energy contained by the enzymes. To prevent the intense power of the charge from overwhelming the process, NADH releases a small amount of its energy at each of the four stages.

When ATP arrives at the site where energy is required, the enzyme ATPase and a molecule of water reduce ATP (triphosphate) to ADP (diphosphate) plus a phosphorus molecule. After it has discharged its energy, ATP is reconstituted. It is changed from diphosphate to triphosphate by compacting ADP with a third phosphorus molecule.

The compacting requires the force of four high-energy electrons because three phosphorus molecules are an unstable structure.

ATP affects the vascular oxygen supply and is necessary for every muscle contraction. It is involved in cells communicating with each other by neurotransmitters in the nervous system. Each molecule of glucose produces about 30 ATP. Glycolysis and the citric acid cycle create 4, and 26 are reconstituted during oxidative phosphorylation. The total quantity of ATP stored in the human body is about 100 grams, but every day we use the weight of our body in ATP, as each ATP molecule is recycled 1,000 to 1,500 times. ATP supplies the body's energy for work, synthesizes RNA, and provides the energy for enzymes to move nutrients into cells and remove wastes and toxins — even against gradients.

Oxygen constantly needs electrons to fill its outer orbit. Oxygen's craving for electrons propels the movement along the electron chain. Half of the oxygen expires as a watery mist at the end. Sixty percent of the energy used to transform glucose to ATP is released into the particle storm as heat and disordered movement, increasing entropy.

## THE RESULT

Altogether, between 75 and 80 percent of our energy is released as disordered heat. Twenty to 25 percent is used kinetically as work to maintain our bodies. Most of the energy we produce by breaking down matter (food) can never be returned to orderly movement or used for work again. Producing entropy is much more efficient in the multicellular engine, which we are, than it was for the initial algae life form.

The story that we are productively focused on work and creation is the wrong side up. The right side up is that our job dissolves matter into energy and dissipates it in heat and waste. The human mechanism is inefficient in creating work. Where we excel is at generating entropy, dissolving gradients, and breaking down matter into dissipated energy.

Since we started as a chemical process to reduce gradients it should not be surprising that that is our purpose. Since we started as an intermediate process of energy between the bound proton and the relativistic photon, centered on the dance around the electron, it is not surprising that we continue to inhabit the electron world. Since we started as a function to fulfill the laws of physics following the direction of the cosmos, it is not surprising that we continue to be obedient to the laws of physics, particularly the second law of thermodynamics. That is what we are. Period.

We claim that this is happening to produce culture and thought, which stands at the apogee of the entire process and is the meaning of human existence, perhaps of all existence. However, all our cultural constructs and social organizations are at the service of chemistry, physics, and the laws of nature. Culture is puzzle-solving. The executive function of the human mind leads us in that direction. Beauty and math are puzzles — the best. They are hand-maidens to entropy as we figure out how to navigate our worlds, harvest matter for energy, break down complex structures into heat, create entropy, and reproduce ourselves as entropy-producing machines to keep everything going. Not only do our bodies primarily create entropy, but the work for which we use our energy and the culture that directs it are primarily devoted to creating entropy. Our species' significance is subsumed under chemistry, which in turn is subsumed under the laws of physics. Look to science for meaning, not to the species.

The idea that evolution aims at anything is anthropomorphism. We are simply burning through matter, and with each evolutionary change, a species does it more efficiently in mutating environments. It is all about the process of deconstructing matter. Evolution does not mean anything other than increasing the efficiency of creating entropy. The evolutionary process is a much larger process, while life is a tiny part. Surmounting adaptive challenges is simply finding a way to continue to convert matter. The human cellular creature's process is identical in function to single-celled or simple multicelled beings. We are better at it individually, although they are better collectively as species because of their numbers.

Every cell of our body contains the direction of the cosmos. Existence is a given, but then we struggle to continue the process. The purpose of any life form is whatever an entity is good at in its process. It is hard as a human to think of our species as having so little importance. The dominant direction of the universe is the breaking down of the gradients, which includes all composite matter, which in turn includes us. In the face of the dominant direction of the universe, our human fate is painful and tragic as we seek order and safety.

Here is the essence of our problem. We think of destructiveness as alien to our true nature. Our destructiveness is the propensity we have for cruelty, greed, war, the sacking of the Earth, the seeking of dominance. On the other hand, we praise finding innovations and increasing our consumption. We make that the backbone of our social and economic life and make it central to our capitalist ideology, which we claim is simply how the world works. If we are to change the trajectory of the human project then we must acknowledge how central entropy is to our purpose, while at the same time it is the bane of our existence. Once again, therapeutically our species needs to acknowledge the unconscious truths governing our behavior and our compulsions. Without the acknowledgement there is no prospect of change.

So, how do we live with tragedy? Palliative care is a therapy that attempts to control the symptoms so that the individual can live as productive a life as possible. It seems that the therapy needed for humanity is palliative care. We have sought this already by using sugar pills of jus-so stories that flatter and deceive us and by pretending that life is a comedy rather than a tragedy, with a happy ending rather than a sorrowful one. Our first step is to seek truth, which is abundantly present in science.

# Chapter 6

## There Is Only the Brain

Objects in our environment reflect light within the visual spectrum of varying lengths of photons' waves. They enter our eyes through the cornea, which focuses it with a lens onto the retina's photoreceptive cells. The cells detect the photons of light and produce neural impulses, which the optic nerve transmits to the ganglia in the brain. The ganglia send the information to the cortex. Simultaneously, other waves or molecules are sensed and together they are lumped into objects. The objects and their movements are recorded in our neurons, which collect them into constructs of our world and our expectations. Neural impulses registering wavelengths and molecules bonding are what form our experience.

## THE STUFF

There are 171 billion cells in the human brain, about half of which are neurons, and half are glia — mainly in the cortex. According to Alan Jasanoff, a quinoa grain-sized piece of cortex contains 50,000 to 100,000 neurons of 100 different types, a few miles of axonal wiring, and a billion synapses. Each neuron communicates with 150 other cells on average (150 times 85 billion connections). Neurotransmitters and neuromodulators chemically convey the electrical charge of the axon between neurons and modulate the charge's intensity. These modulations increase the combinations and the degrees of experiential intensities. An estimated 100 trillion synapses between axons and dendrites convey information between neurons and construct our experience. What is most important is the variety of the 100 trillion connections between neurons. A thoughtful perusal of these vast connections shows us that our brain connections sufficiently account for our experience. We do not need to resort to any other entity, such as the mind.

The networks of connections and layers of interacting information stored within the brain are human experiences and functions. Our experience is structured by the hierarchy of neurons in the brain, feeding forward and looping back, connecting modules with different specializations. Neurons loop backward to the beginnings of the hierarchical sorting to allow stored experiences to influence and interpret incoming signals. What has been shaped and stored influences our experience and responses. Olaf Sporns has shown that any given neuron is only a few connections away from most other neurons — an astonishing fact, given the number of neurons. Even though the spiking is slow — axons fire at ten to twenty spikes per second — the combination of so many closely connected neurons makes for very rapid calculations — tens of millions of signals in and tens of millions out, each second.

The 1,000 or more types of cells in the brain, the various kinds of neurons, the very different neurotransmitters, and the neuromodulators that affect emotion (a source of motivation and decision-making) add complexity. Inputs from the body to the brain include the vagus nerve connecting organs and the bidirectional enteric nervous system linked to the gut. The enteric system contains over 100 million neurons, and the gut contains virtually an infinity of bacteria that influence the brain.

A storm of particles inside and out — waves and molecules — come in through the senses, and waves and particles, in turn, interact with the environment. But what else is inside your skull besides this storm? Particles and waves do not produce separate entities that are phenomena and ideas. They *are* the phenomena and ideas. What else need you ask for?

The brain creates plans by connecting the neurons without our awareness. They are our feelings, thoughts, and memories. Perceptions and thoughts sum up and reflect on our memories out of the previous connections. Neurotransmitters and neuromodulators shape our feelings. When we are aware of our intentions it is without our knowing the brain's unconscious sorting. Different brain parts loop together, and parallel processes sort our experiences and initiate our actions.

## THE PROCESS

Perception is influenced not just by the information that comes from the sense organs but also by what the brains decide to do with that information, i.e. what they can take out of the information that is useful. We see what we need. This is a general principle of cognition.

Neuronal processing rests on sorting and repetition. The process sorts the environment according to learned experiences, primarily fear and greed, and then proceeds to approach or avoid. Feedback from higher states in the sorting process modifies the earlier sorting of what perceptions are vital opportunities or threats. We learn from later results how to modify our judgments of the initial inputs. The brain's process is reflexive.

Every second, we activate tens of millions of neurons in response to light and sound waves, the molecular chemical stimulus of tastes and odors, and skin pressure. The stimuli transform into electrical/chemical charges of neurons with potentials for actions. A sufficient charge triggers chemical neurotransmitters to cross the gap between the synapses from axons to neurons, creating a new action potential in the next neuron. The process is hierarchical. Collections of neurons sum up the strength of their charges and pass it up to the next level. Eventually, an action, decision, or thought is initiated.

When conflicts occur at the highest levels among the hierarchies, action is determined by weighted neuronal summaries sparking the electrical spiking. There is little or no freedom. Our conscious experience is identical to the final discharge from a small portion of our neurons.

The neurons' patterns reflect the interactions between us and our environment. We may become aware of our unconscious shifting perceptions and potential movements when we jump back and forth until a decision is made. Biology wrote our purpose. Our bio-dynamic is our reality, not our pictures of ourselves as conscious and in control. We are a mechanism fulfilling a task.

Unconscious processes form the vast bulk of our awareness and response to our environment, a fact of which we are rarely and only subliminally aware. For some of our plans and movements, the unconscious becomes conscious. We refer to unconscious competence in controlling activities such as playing the piano or doing an operation. It is spooky how we trust our unconscious processes even though we fear they will fail. It is as if we are controlled and are an automaton. Perhaps we are.

## THE PURPOSE

Our brain senses our environment and triggers movement in response. The prefrontal cortex identifies environmental patterns and the anomalies and novelties that signal opportunities to acquire energy, or dangers of being acquired by other humans or other species for their energy requirements. The PFC privileges social cooperation, structures hierarchies, enhances language, and maximizes our skills to collect and manipulate matter and convert it to energy.

We are so focused on energy and avoiding becoming energy for someone else that we only know the conclusion of our experience and not the process. We think the process and the conclusion are two separate objects rather than a single processing of information. We are so imbued with this dualism that even though we may understand how photons become transcribed chemically to become information electrically and are embedded in the neurons, which triggers movements, we retain the separation of mind and body. The outside experience of extension and externalization and the brain sorting sensations and actions are on the same ontological plane. There is only one world, not two, and we are totally in it.

Even in decision-making, when we are alert and attending to alternatives, we listen and weigh what is not conscious, going back and forth, feeling for the answer to "come." We can vaguely feel the unconscious working in the brain. Similarly, when we try to find a memory, we wait until it bobs up like an object from the turbulence at the spillover of a dam.

It is the weighting of the neurons that makes the decisions. Decisions are the outcome of probabilities unconsciously calculated from interpretations of given and genetically privileged experiences, our genetically conditioned responses, and chance environmental formations. The point is that almost all of this is unconscious and selectively delivered for executive action, which is when we become aware.

Neurons operate by Bayesian logic, going from the best belief to the next, encoded in the neurons' electrical action potentials. They will be wrong sometimes and correct other times. From the outside, a good or a wrong decision is made, which gives us the notion that there is free will. Based on this, we describe ourselves as willful and responsible. Based on that description, we construct ideas of our morality, mainly as it affects how we cooperate and offer protection to each other.

## WHO ARE WE?

With sixty trillion cells in communities (including the gut bacteria), communicating, pulsing, writhing, flushing, pushing, and exchanging, why exactly do we think we are in control? By "we," I mean the individual who thinks it is doing it. We did not create sixty trillion cells in their relationships and processes. They created us. So, the question becomes, why did they create us?

Our nervous system responds only to relevant aspects of our environment. What is relevant? Unlike a bird or a bat, we do not seek out insects. Unlike a tree, we are not geared primarily to photosynthesis. What are we geared for? Like some other species, we are social animals who organize hierarchically. What we do together cannot be done as well singly. Our more intensely wired brain has given us our unequalled ability to burn other species as food/fuel and increase entropy.

The human brain has dramatically facilitated our species' ability to convert matter to radiation. The most significant step was not only burning matter internally within our body but also burning matter externally. Fire enabled us to process more calories internally by breaking down our food before we ingest it. Our brain demands a large amount of energy to function, and the breakdown of food through fire enabled substantial growth, preparing it for further triumphs. The big step by our species was gaining the ability to burn our environment, which empowered us and increased our production of entropy. In the last 250 years, this ability has been compounded with the move from burning trees to burning fossils. Therefore, we were empowered to enter a new epoch, irrevocably altering the planet. The evolution of the brain has primarily served the production of entropy.

## PRESERVING THE SELF

In conjunction with the storm of particles inside and outside of us, there is a storm of information inside and outside. The information is categorized and assembled with its conclusions and consequences according to neuronal weights. The *self* is the direction of the assembled choices. We are the information embodied in us that interacts with new information provided by our sensations and communicated to fellow members of our species. Our functioning is built on the patterns we inherited from other animals of converting incoming stimuli to outgoing actions, to form conscious and unconscious thoughts/algorithms, and to acquire the particles and information that construct our awareness of the environment.

Communication has existed among living beings from the very earliest times. It is egregious to describe language as only human. Other species respond to their environment and are aware of their kind, but we hesitate to use the word conscious without motility and a nervous system. The language between plants consists of volatile organic compounds that signal developing threats and alert other plants to perform defensive actions. This form of communication is language — the communication of understanding and responses — even though not verbal. Communication does not have to be verbal to be language.

The interactions of a species with the world it senses are biologically coded information. Information is as much a reality in the material world as energy or motion because it describes orderliness, which is inherent to nature. Language conveys information biologically within the local species. The biologically coded information informs the behavior, beliefs, and culture shared within a group.

Algorithms govern animals. We are programmed to do what we do. Animal minds interact with their environments to process matter, resolve gradients, survive, and replicate. Most animal and human behaviors are unthinking responses — simply acting out the plan. Reflection consists of optimized neurons coming to the fore and fading as their information is weighed to establish primacy. We are animals, but our neocortex expanded how much information we can juggle, increasing our flexibility and predictive abilities.

Brains map together in synchronicity in our cooperative sociality. Shared experiences bias shared brain patterns and create shared meaning. Shared brain architecture and environmental interpretations make for contagious thoughts. Physical connections of neurons in one mind are mirrored by those in another. Contagion in communication by neuronal patterns, fully embedded in matter, controls what we call the mind or the self.

There is no center, only habits of actions and mental approaches plus continuity. The habits are downloaded from the environment or developed on some basis of trial and error. Some habits or beliefs are sufficiently strong that they appear as character and are seen as a core, or a self.

The connections of decision-making neurons adding up to something, relating to other conglomerations, create a temporary unity, which we see as permanent, but it does not last. We become different persons as our environment changes, modifying our conclusions and altering our development. When the summation is different, everything adds up to something other than what it did before. Instead of a self, I call us a semi-permanent summary of a neuronally inscribed algorithm. The self is the sum based on the previous summations in response to present environmental challenges and the calculations of energy needed for self-preservation.

Some actions run so far toward entropy that they counter the species' interest, even if they fit the environment in which we exist. These are labelled morally wrong. For example, contradictions take place if one is violent but lives in a murderous environment, in which case one might fit into a gang or a war party. One is neither credited nor discredited but only lucky or unlucky in what one has inherited or learned.

We attempt to freeze the sum of sums rather than respond to the changes that happened internally and externally. The reason for doing this is to counter entropy. We cling to our beliefs to prevent changes in the summing of sums when we perceive that change will produce more entropy. We have vainly hoped to create an image of immortality if our frozen belief or our sum of our sums lives on after our death in some form.

One half of the brain is devoted to finding the patterns we expect, and one half has evolved to alert us to the unexpected, which may be dangerous. Frozen beliefs — a meme — are a collective summation that aims to maintain continuity, or, more important, predictability. They are lazy ways of quitting, cutting short summations, and not leaving oneself open to change.

To maintain predictability, we interpret experience to fit the memes rather than continuing to sum the experiences when we have reached our limit. The attempt to be static in a changing world is to maintain predictability as well as to favor potential over kinetic energy.

The sum of sums is a changing self with discontinuity in the narrative of an individual. We are alarmed at this discontinuity and strive to reverse the changes by living in various delusions, beliefs, cultural memes, and lies. These provide an illusion of continuity when there is discontinuity in the sums of sums. Neuronal connections come to different conclusions at different times and represent different selves. It is akin to the breakdown of the boundary of the cell, where the cell is the membrane (skin) of the human.

These are the two primary forms of existence: energy and information. Information is based on the forms of changing matter subject to entropy, breaking down. Therefore, the sum of sums is constantly changing. Our concern about our identity is a defense against entropy to stop the flux of information that is summed up to create a self. Our resistance to change fixes our beliefs, cultural shibboleths, and memes, but nevertheless the sum of sums is constantly modified. This raises the question of who that "me" was. The fixing of identity by the resistance of informational change complements our defense of favoring potential over kinetic energy — of which more in a later chapter.

We accomplished little in securing reliable information until the scientific revolution, 250 years ago, introduced novel instruments for perceiving and a mathematical language to interpret what was discovered. A different epistemic was born, using different sensations to access and examine different worlds with objectivity and verification. With the scientific revolution, the possibility of truth existed instead of only contagion, and much stronger beliefs powered the sorting of the brain.

In summary, the brain's functions evolved to facilitate the universe's evolution. Our brain highlights how to process matter into energy, avoid predators, and reproduce. We are wired to recognize the opportunities afforded to us by our environment. The phenomena we sense are shaped by our nervous system, as are the actions responsive to our needs. We receive input from the environment through our senses and give output to the environment through our movements. We are embedded in matter and energy moving around and through us from our environment. The flow-through is the process of our existence. The brain builds the self. "I am" is stored in the patterns of connections between brain cells. The death of the brain is the death of my mind. My mind is my brain.

We are molecular, a tiny cone of space in movement at a point in space and time, which is singular. We name this a personal experience and give it a separate identity. We are a vortex of dust in a storm, then the storm settles back and slows down.

Our subjective world is in unconscious flux. We attempt to maintain an orderly permanence mentally as well as physically. We must modify our beliefs and, therefore, modify the sorting of the perceptions instead of striving to gain greater security by fixing habits in place to create immediate responses. Habits, thoughts, and actions are efficient until the world changes. Our capacity to change the neuronal weights that are our beliefs may enable us to survive.

What this is to an individual is the same to the species. We are embedded in our beliefs, many of which go back to the very beginnings of our civilization. Some are false. The world of the last 200 years has been in considerable flux, requiring swift changes in our habitual ways of thinking. Adaptation will become even more important as we learn to change our genetic inheritance and incorporate artificial intelligence. Simultaneously, we are endangered by the Earth's destruction and the growth of our species' numbers. So, we are in a push-pull around beliefs about whether to cling to what we have or modify them. This has created unbearable stress. We need new algorithms. Fortunately, we now have science, which puts belief on a firmer foundation in a statistically credible reality.

# Chapter 7

## The Fallacies That Create Hubris

Seven delusions separate us from nature. It is obvious to most of us that our core is not matter. We came from somewhere else and are going somewhere else. Akin to this is the belief that God appointed us the Lords of the Earth. Even those who are not spiritual have an abiding conviction that we are entitled to dominate the Earth and all other living things because we are superior to nature and its inhabitants; the Earth was placed here for our use. Even those not committed to supernatural beliefs think that our minds and cultures exist ontologically apart from the laws of physics. Although the skim on which we live is superficial, the enormous population increase reveals our optimism that we have fundamentally limitless resources. Even in the face of scarcity, we are convinced that our limitless ingenuity will save us through science and further developments such as artificial intelligence. Even though we are individually fragile, we feel safer within our identity groups and institutions. We are confident in the survival of power and status structures, economic systems, religious values, and national identities. We are good and deserve a happy ending.

## SEVEN FALLACIES

**We believe we are aliens.** We are not matter in our essence. Through luck or hard work, some or all of us will continue to live after death in heaven, hell, paradise, in a nirvana state, or be reborn. Life is a comedy, not a tragedy, and it ends happily — for some of us.

Belief in the existence of a separate world is widespread, almost universal. Some even welcome an anticipated apocalypse on Earth that will transform the planet into eternity. We reluctantly anticipate our demise as individuals, but our actions reveal our belief that we will survive "somehow."

The purpose of faith is to rescue us from entropy. Time is associated with the Earth, while eternity is a separate logical entity. Our divine destiny rescues us from our fragility. It is delusional to believe that in the entire cosmos, a few creatures who live on a speck of a speck of rock have a different substance than everything else.

**We are Lords of the Earth.** Scriptures and creation myths reinforce this belief. Our culture is separated from nature. As Lords of the Earth, we have the right to sack the world. This belief reinforces the centrality of corruption and psychopathy within social organizations. Even someone as ecologically aware as Pope Francis recently reaffirmed the

Christian belief that our progress stems from the God-given entitlement to dominate nature.

In fact, we are *children* of the Earth. Our skin is a membrane, and our body is 60 percent water. We carry around the ocean from which we came. Oxygen makes up 21 percent of the Earth's atmosphere, almost half of the crust (in the form of oxides), and two-thirds of the human body as a gas and in compounds, principally water.

Carbon is 18.5 percent of the human body. Because it has four electrons that can be shared with other atoms to create molecular bonds, carbon is the largest source of compounds. Carbon creates life's complexity. The cycling of carbon throughout the biosphere underlies the existence of all life. Radiation from the Sun and the Earth's core heat our bodies. The ingredients of the crust of the Earth fully constitute the human body.

Plants absorb photons through chlorophyll in their membrane, and the kinetic energy of photons carrying the electromagnetic field is transformed into chemical energy by an electron chain movement. Water is separated into hydrogen, electrons, and oxygen. The oxygen is released into the atmosphere. Carbon is separated from the carbon dioxide that is ubiquitous in the atmosphere. The chemical energy consolidated in molecules synthesizes carbohydrates out of carbon and hydrogen. Carbohydrates — stored in the plants' cellulose — are transformed into glucose by the animal's digestion to fulfill basic energy needs. We are totally integrated with the Earth rather than existing apart as its Lords.

**We act as if our resources are unlimited.** Our population is eight billion and will reach ten billion later this century. Although our conception of the world has shrunk with the ease of travel and international communication, we believe the Earth's bounty is ultimately limitless. We have a lot now and are confident that we can figure out any difficulties we come up against and surmount the scarcities we face. We are manic with abundance, having unlocked limitless ingenuity. We float high on our appraisal of resources that defy the limits of the thin crust we live on.

Compared to the cosmos, our rock is like a grain of sand in the Sahara Desert. We live in a tiny space floating on the skim of an insignificant rock. The Earth, with a diameter of 12,700 kilometers, contrasts with a diameter of 930 *trillion* kilometers for the observable universe. The Earth's iron inner core, with a radius of 1,200 kilometers, is surrounded by a liquid outer core 2,200 kilometers deep. Closer to the surface, the mantle has an average thickness of 2,900 kilometers. The crust of the Earth is merely 30 km thick under the continents and 5 km deep under the oceans. The crust floats like a skim on the plasticity of the mantle. The liquid core pierces the crust at fault lines between the floating continents.

We live on top of the crust. The deepest point in the ocean is 11 km down, and the highest mountain is almost 9 km above sea level. The troposphere is 1 km above the highest mountain. Life exists within 21 km of the surface, compared to the 6,300 km to the Earth's center. Our habitable area on the surface is much less: we are squished into about one-third of it. Water covers the rest. The surface we live on is almost flat, much less than 21 km in height. Our living space has been compared to the thinness of the skin on an apple.

**We assume the Earth's continuity.** Our historical memory is weak, and we do not identify with how hard life has been in previous generations. They did what they had to do to last, eat, and have some pleasure, which was good enough. We know from science that vast changes in the world's climate and structure have occurred, but we are divorced from any direct experience. It could not happen to us.

The skim of the Earth has experienced extraordinarily violent changes and will do so again. Climate change is causing oxygen levels in the open ocean and coastal waters to decline by roughly 2 percent. Hypoxic dead zones have gone from 45 in the 1960s to at least 700. Some encompass thousands of square miles. Our species is so recent that our knowledge is like two frames during an entire movie; climate change faces us with a third discontinuous frame, and — as happens in films — it is a shock! The world is imperiled.

**We believe our minds and culture are separate from the rest of the world.** We believe our culture guarantees us endless progress — even immortality. We believe we ontologically exist apart from all other living species and the laws of the universe described by physics. By separating ourselves from nature, we became convinced of our superiority over all other life forms.

Species' cultures change as the environment changes. We do not stand apart and above, for we have inherited the evolved characteristics of prior species. These accomplishments include communication, personal display, social organization, dance, toolmaking, caring for the young and injured, adoption, warfare, building shelters, recognition of individuals, songs, and teaching. Genetic accidents that enhanced the endeavors of earlier species as they sought energy in their niches built our species' culture. Culture is an infrastructure that enables us to find the molecules we need chemically to break down complex objects, undo gradients, and create entropy and waste. Culture is a direct result of the laws of physics and chemical processes. Culture serves the laws of nature and does not exist apart from them.

**We can live robustly indefinitely.** We deny change and fragility. We are confident that our power and status structures, economic systems, religious values, identity groups, institutions, and nations will survive to guarantee our individual and collective safety and our species' immortality. If worse comes to worst, our intelligence and science will save us. We believe that our ingenuity is limitless. If we cannot save this species, we will create another one. We believe we are endlessly exceptional to the point of becoming immortal as a species.

Our sketchy written history extends a mere few thousand years. Compared to the universe's life, the human's eighty-year life is analogous to a virtual particle emerging from and returning to its field. We are ephemeral and quickly forgotten. No matter how much continuity or even immortality we hope for, human life ends within an infinitesimal fraction of the lifetime of the cosmos. We live within a narrow band of knowledge and experience with an emotional illusion of permanence.

**Lastly, we are good, and therefore life should turn out well for us individually and collectively.** I will address this delusion in Part Two of this book, which comes up next.

## CAN WE RECOVER?

These seven fallacies persuade us that we have greatness and security. We live large and are more significant than anything. Factually, we are a fragile, contingent species, and the existence of each individual and even the species is tenuous. Realistically we face the possible decimation of our species if we turn the skim we live on into scum.

Our belief in permanence is an artifice. We push away death in all the areas of our existence, but it lurks within our minds, as it must because of entropy. We are uneasy despite our illusions.

Would the species be called psychotic if it were judged as an individual? In earlier chapters, I focused on how the cosmos works, but in this chapter, I extrapolate from the previous ones how we interpret our world. It is not just how we think about it; we mostly acknowledge that science is telling us truth. It is more about what we believe and how we act based upon our habitual beliefs, which are not anchored in science but are phenomenologically ancient. The beliefs that govern our species' situation are bizarre.

Therapeutically, how likely are we to be able to change our way of thinking about our place in the universe, on the planet, and with each other? To become able to change from a psychotic state requires a breakdown. The threat becomes existential. If it is an individual, he or she ceases to function productively in relationships with others, and therefore becomes cut off and trapped within himself. Is this what is required for humans to become amenable to abandoning their delusions?

There is also the threat, as in the risk to an individual, that the potential for change is left too late. There are also signs that a substantial percentage of the population would prefer to follow their leaders into doubling down on their delusions and turn violently against rationality.

In this century, there now are threats that are existential for the species because with nuclear warfare, climate change, and overcrowding, we are degrading the base on which we live. How far would this have to go in terms of environmental destruction or nuclear war? The species is in a breakdown, but climate change is a slow-moving one, which has allowed us to rationalize. Nuclear war, potentially initiated by a middle level state, could rally the rest of the world to more rationality. At some point, either of these events could be a sufficient shock to motivate humanity toward substantial change.

In the case of an individual, there are mechanisms within the society that attempt to prevent harm. However, for the species, no external force exists that can confront its irrationality and prevent harm. Even if there is some intervention by the force of combined states, there is unlikely to be a complete recovery and a vigorous acceptance of reality. As with an individual, the chances are that even in the wake-up call of a collapse, the species might improve but would remain wounded.

# PART TWO

## Lords of the Earth: Why Are We Immoral?

Part Two explores the consequences of our belief in our human exceptionalism and separation from nature for our relationships with each other. It examines the epistemological consequences of our sensory limitations and the ambiguity of our verbal communications. These limitations facilitate our myths as we attempt to carve out some permanent identity, denying the degree to which we are part of processes beyond our control, and in defiance of a world of flux. Part Two derives our moral behavior from physical laws, particularly in the struggle to maintain potential energy and prevent its decay into kinetic energy. The dominance of greed and cruelty is grounded in physics. Greed and cruelty spring from our core as chemical replicating creatures serving the increase in entropy and destruction of gradients in temperature and density.

Firstly, we are story-driven fabricators. Our behavior is restricted and controlled by the categories of stories that frame our meaning and circumscribe our behavior. There may be laws for stories. Secondly, our behavior is conserved from the strategies of previous species devised by trial and error to avert danger and forestall entropy. Lastly, our behavior is determined by the laws of physics, particularly the dance between potential and kinetic energy. The accumulation of potential energy is privileged by the human species as we try to survive long enough to complete our evolved chemical process.

# Chapter 8

## Our Ambiguous Storytelling

*This book is a story.* It describes our places and purpose in the universe. It does not include equations and does not make verifiable predictions. It is an interpretation of science, not science itself. This book puts forward entropy as the force that shapes every aspect of human experience, focusing on the purpose of human existence.

## TWO LANGUAGES

There are only facts and stories. Facts are observations of matter aided by mathematics. Hypotheses can be checked against observations and verified by experts. Stories describe nature and people using our language of intentions and purpose. The human viewpoint is central to stories, and our significance is our primary interest. We want to figure ourselves out. We are both subject and object.

Information expressed primarily in words and mathematics applied to the analysis of matter have unequal probabilities. Repeated experiments test mathematical predictions. However, with the advent of big data and the developing inclusiveness of algorithms, human interactions are being formulated mathematically with predictive probabilities.

We are most comfortable communicating verbally and linking our personal experiences using signs we share with others. We tacitly assume that a shared sign maps a shared experience. We believe we can guide our actions with a map of words that brings us to agreement. However, our vaguely understood words only loosely relate chunks of experiences that vary from one individual to another. The clumps of associations are ambiguous and vary from person to person and culture to culture. Our expectations have a middling probability for being predictable in contrast to the certainty, or near certainty, of information described by mathematics and objectively compared in experiments or with probabilities inferred from large amounts of data using artificial intelligence. Although the evaluations and information about our experience exchanged in words are essential, the results are vague and repetitive. In contrast, the story of nature expressed mathematically and in algorithms is vast, cumulative, and approaches certainty.

## WE ARE OUR STORIES

Initially, we only had our own story, but we knew there was more because we were subject to the vicissitudes of natural forces. Forceful winds blew up unexpectedly; lights moved in the sky, the sun by day, the moon and stars at night. Struck flints created sparks and fire. We interpreted nature as an extension of human experience because that was the only story we knew. We personalized natural forces as a society of humans with greater powers. The narrowness and ambiguity of our single species' experience skewed our interpretation of nature until science took hold.

Our brain is a predictive machine that seeks energy and avoids dangers. Our fear, a product of our fragility and uncertainties, drives us relentlessly to manufacture explanatory stories. Stories include gossip, conversational hearsay, the news, novels, scholarly inquiry, and everything in between. Both "factual" and imaginary stories probe the mystery of human interactions and attempt to determine what is significant. We are endlessly interested in each other because we are dangerous. That no one is safe is the essence of our desire for meaning. "What does the other see and where? What is it? Can I do it better and faster?"

Our world is mediated through stories and shaped by social feedback. The structure of our lives is a web of stories that we collectively create and individually adopt. Our meaning embodied in stories controls our affective, political, moral, intimate, social, financial, and status lives. We are immersed in stories more than in information. Stories are our fundamental reality.

There is truth, but tales are far more numerous. Facts inform, but we have many more stories about the facts than there are significant facts. Tales give endless variety, whereas scientific truths are singular statements that stick to observations. The variety of stories is not harmless. The proliferation of stories, opinions, and lies moves us toward randomness (as does entropy) and destroys focus, work, and principle. There is an unrelenting push to disorder and vacuity, which hopefully will be redeemed and replaced by reliance on science, data analysis, and artificial intelligence.

Unless we are connected to a story, we suffer psychological and affective death. We need stories to feel safe. We are desperate to maintain our story/identity and to link our stories to the stories of others. Being included in a powerful group's point of view is necessary for our existence. Being sidelined is unsafe. We generate security by mirroring others and enhancing dependency. We create imaginative structures of shared belief and worship. We share emotional, intellectual, and imaginative stories like atoms creating bonds by passing electrons back and forth.

We are biologically determined to maintain the strongest bonds with our immediate kin group, and as the human population increased, kinship bonds extended to clans. Separation from the clan has often meant death.

We are embedded in the safety net of our immediate culture's beliefs. Despite our attempt for salvation from uncertainty, we are still subject to acts of terrible violence that destroy kin, clans, and beliefs. Stories of historical injuries and insults by other groups, tales of real or imaginary heroes and conquests, and memories of the ebb and flow of property and territorial boundaries bond members of society and fuel resentment, heroism, and aggression. Collective worship and repetitive rituals give life meaning intellectually and imaginatively and foster the hope of acquiring immortality to overcome entropy.

Making believable stories individually and collectively is central to our success. Securing individual and family safety is our central motivation. Stories about family, love, and religion emphasize commitment. However, they are in the minority. In contrast, we construct heroic stories to amplify our unconstrained exceptionalism as individuals or as a tribe. For those who privilege unrestricted growth and heroic achievement, mere safety for the average person, let alone the marginalized, is viewed with contempt. Stories of personal exceptionalism and celebrity beguile. It is doubtful that the enthusiasm and hubris connected to collective and personal lack of constraint can be overcome any more than one can slow the sun's burning.

In Plato's prophetic description, we are compared to captives in a meagre cave of restricted sensations who become better and better at discriminating nuances among the shadows of objects but cannot observe reality. We poke around in a conversation, bewildered, cataloguing the contents of our mysterious world. We grope and grasp and call it culture — of which we are exorbitantly proud. Just as closely packed atoms jiggle and create heat, which then dissipates, closely packed humans jiggle and create the heat of dissipating culture and entertainment.

To use a water analogy: Many shallow rivers flow through everyone's brain, carrying ideas from and for all our fellow humans. Occasionally, an individual climbs out of the flow onto dryland and says, "Here is another world. We do not have to be swept away." However, the agreement is that we float down prescribed rivers, and no one questions the direction even though the river is shallow, can convey little, and makes little sense of the direction. Agreement with our group is a compelling and universal human trait. Even the informed and intelligent feel that they must believe and embellish the imaginative cultural stories. In this world, mathematics and science are our respite, comparable to observing Plato's Forms.

Art is an exercise in self-preservation, as we are thrilled by our discriminations between what is predicted and what is unexpected. Our brain is hardwired to discriminate between what is habitual, expected, and safe — for which we have predictable responses — and unexpected stimuli and actions. The essence of art is the combination of the expected and the novel. The individual artist's work is the repetition of content with slight variations. Significant variations occur between artists over long periods.

Stories also transmit norms of behavior. Stories with many variations include tales of intertribal warfare, vengeance, seduction and loss, power and wealth, teleology, supernatural beliefs, jockeying for status, and cooperative values. We find gripping stories about tracking predators and defending against them. Many stories emphasize in-group exceptionalism and the markers between groups. Other forms of storytelling include dance, repetition and ritual, gossip for tracking and knowing others, honor stories, the worship of ancestors, traditions, stories of gods and immortality, and tales of identification with other species. In the earliest times, storytelling conveyed from generation-to-generation practical information about fauna and flora, weather, local geography, and neighboring groups. Singing and dancing created group cooperation.

Storytelling is the core of communication and identity, which holds society and individuals together. There is little reality in our world concepts beyond the story, and the story is the reality. Stories are an act of imagination loosely related to social realities but are based upon or pastiches of past fiction. The mythology of human life is that the stories are real. We can, however, see the arbitrariness of fiction in stories deemed to be true in their contradictions, particularly stories told at different times in history, in various cultures, and by individuals variously describing the same events while believing their descriptions are factual. My guess is sufficient data and analysis will discover the laws for the conservation of narratives. Older stories are simply modified from new points of view but with similar narratives and outcomes. The accumulated novelty, originality, and wisdom are slight.

## STORIES' LIMITS

We are *artifices* assembled by our shared, locally contrived, and endlessly repeated stories. Our beliefs and goals bond us to fictional entities until we become fictional. Universalizing our local beliefs and naturalizing our cultural conventions transform us. Because our survival depends on our collective stories, we aggressively defend their truth.

The human project, expressed only in words, has no hope for a secure foundation. We attempt to cull a sensible narrative from nothingness. We hope to preserve our narratives as a solid world on which we can stand, assert ourselves, and believe that this is "just so." However, we are quantum and discontinuous, as matter is. We could be any and every possibility, or no possibility.

Socially, stories are essential to our mind reading when we attempt to overcome the ambiguity and emptiness of our knowledge of others by pressing a concatenation of our senses into guesswork to create an imaginary picture of what exists in other peoples' minds. Opaqueness applies equally to our own minds during self-reflection. Stories approximate human thoughts and behavior, but they are inexact. Stories attempt to describe our subjective human experiences truthfully, but in their totality, they fail.

Lying to oneself is also a story. One can know the story is not the entire truth, and one may even know the truth, but what makes the lie story plausible? Is life the same as a play where we suspend our disbelief? Rationalization enables cruelty. In Nazi Germany, Goebbels facilitated a "knowing without knowing" as long as no one spoke out. Goebbels was clear about not rubbing atrocities in the faces of the German citizens. He expected they would prefer to feel conflicted in silence, or they would rationalize what was taking place and remain passive. The populace feared social isolation if one spoke of the murdering of Jews, although it included women, children, the disabled, and the elderly. Families silenced criticism among themselves. In another example, torture was recently redefined as intensive interrogation in the United States. Consequently, most citizens put their safety first, morally compromised themselves, and pretended that waterboarding someone was not torture.

How many stories exist that people live by but know they are at best uncertain and probably fabrications? Most humans are firmly in the grip of local thinking and values. How can anyone escape from them? What wormhole can lead us into outside thoughts or values?

## THE MEANING STORY

"Meaning" is a story that humans tell. The meaning story relates us to each other and everything nonhuman to us at the center. The meaning story is how we organize experience. We construct narratives around two fundamental questions: What are we individually and collectively, and what is the purpose and end of the world?

Our stories about meaning include purposes and goals and privilege the significance of endings. Historically these stories have placed us above nature and confirmed us as the purpose for the universe's existence. However, they have little to do with the substance of the world outside of our society. It has no independent existence. Even intelligence does not enable us to escape from the hubris of our species' point of view.

Our meaning story brings us comfort and some safety because each human moment and interaction inherently lacks clarity. Behind everything, there is a tragedy. Many possibilities exist, possibly even an infinite number of causes for actions, or perhaps there are none. Actions appear random, proceeding from nothing. The opacity of ourselves and our interactions entails loneliness, disorientation, and uncertainty.

We are randomly born into circumstances that are sometimes good and sometimes awful. We may be given an affluent or privileged life, a poor and frightened life such as in a war zone, a life where it takes menial or horrendous tasks to survive, or one that is cursed with disabling diseases. We all share the pain and brevity of life. Our circumstances, the internal bias of our nervous system, and our genetic inheritance primarily preordain each life's outcomes.

In defiance of our reality, our story continues to claim that we have free will. We choose, make sense of our circumstances, and control how our life unfolds. However, that is just a story. In stories, we seek safety and comfort, the illusion of a free future, and practical skills in manipulating objects and people. Stories are critical, bringing comfort in a world of unspeakable pain.

"Just-so stories" are the foundation on which we find commonality and humanness. Yet each observation of each mind is from a different point of view. We are up against uncertainty and ambiguity at the heart of observations. There is no assured measuring stick. There is no close approximation to truth in the human narrative — not even a whiff of certainty.

Force fields of narration link us into small communities during our short lives, pinning us down. In reality, there is no story. We process in and out of consciousness and on to nothingness. Our individual opaque lives and characters come and go out of existence like virtual particles. Human stories are for humans what aromas in the air are for other animals — they orient. Individual stories are fragile like the aromas; the slightest breeze and they are gone! Is the individual human life just a story?

Art exploits ambiguity to open diverse and contrary possibilities and reveal new information. However, most stories are not art; they close exploration down rather than open it up. Most stories merely deal with emotion and fantasy, even though we hope that putting experiences into words will bring knowledge and enable prediction to increase our safety.

Stories about the meaning of human existence embed a fantasy about our ability to control threats and create safety. A governing story at this point in the scientific revolution is that humans can save themselves through knowledge. Science is part of the story of control, and we hope there is a different direction than history and society becoming unraveled by entropy. Science is the biggest fairy tale because it accurately describes reality but gives false hope. Science, embraced as a salvation story, serves the same purpose as religion, narrative, or art.

The world of our senses, the cultural world of our interactions, our interpersonal relationships and psychology, our literature, poetry, and beliefs do not reflect the actual workings of the world. When we ask "What are we?" it appears that we are a fabrication in stories, and our truth lies elsewhere in mathematics and science. How is it that the world and mathematics mirror each other?

And yet, "story" is our soul. We locate ourselves in the world through our language and within fiction. The environment, evolution, and our genetic heritage shape our stories. Explanations and habits frequently not based on truth or reality shape our small species, barely existing on the crust of a tiny planet. The fabrication, the analysis of stories and meaning, and the responses we make to them are necessary since it is the structure within which we must live, relate, seek safety, and survive unless we can shift to a new modality and allow our lives and societies to be based on and governed by scientific and mathematical information. Everything, including human behavior and storytelling, is matter in motion.

This book is a story resting on incomplete human knowledge, seeking the point of view of a mind looking at the cosmos from the outside. A more accurate story could be told with the mathematics of probability to back it up. I hope that someday, artificial intelligence, together with sufficient data about humans, their behavior, and the laws of physics, will be able to make better sense of the place of human beings in this world.

Psychotherapy for an individual is sometimes a conversation with molecular manipulation, primarily of the brain. The attempts made by philosophers, theologians, and political theoreticians to persuade the species to adjust its goals and the means to achieve them have had only marginal success. What is proposed here is that therapy for the species requires a different language, addressing a different reality. What is not touched on in this chapter but will be addressed later is also the possibility of more advanced molecular manipulations of the mind and body. Much greater mathematical and scientific literacy among the total human species, male and female, is necessary, as well as language expansion through the algorithms and mathematical notations of artificial intelligence. This expansion is happening, but will it be in time?

# Chapter 9

## Cellular Behavior Is Conserved

Behavior is preserved from species to species and over time. There are profound resemblances in the behaviors of replicating chemical and cellular creatures from the earliest times to now. All species' behavior patterns are stereotypical actions designed to maintain individual safety and permit biological and chemical functions to produce entropy in a particular environment.

The behaviors of bacterial species and algae foreshadow our social interactions. Bacteria operate independently, but when faced with predators or an absence of food, they cooperate and support those who are most like them, particularly their relatives. Freeloaders are isolated and left to die. When faced with a predator, a community forms a wall. We behave similarly. We cooperate in the face of difficulties and support the group's survival, attempt to isolate the parasites among us and favor our relatives, whom we expect to support us. We also build walls.

Slime molds live as disorganized individual amoebas feeding and dividing on their own. When threatened by starvation, up to one million cells coalesce into a mushroom-like tower to form multicellular fruiting bodies. Twenty percent of them create a stalk, sacrificing themselves so the rest can move to the top of the structure and form spores, which can last for months without food. Ultimately, water and wind disperse the spores to new and potentially more nutrient-rich environments. The species and even the community continues.

The molds' social behavior is cooperative, but 30 percent remain as loners, exceeding the number of cells in the stalk. Loner behavior is a heritable trait, but the number is influenced by diffused chemical signals that impede aggregation. Loner cells are insurance that the slime mold can regenerate the population. Starving cells send chemical signals to their neighbors. When enough cells sound the alarm, the aggregation process begins. As they aggregate, there are fewer to signal starvation, and loners increasingly remain. Biologist Corina Tarnita posits that loners are a hedge against the possibility that the aggregate body could be eaten by a predator or be overrun by cheaters exploiting the collective. Further, if nutrients return, the amoebas cannot reverse the aggregation process to access the food; however, 30 percent of the community remains to profit from the renewed situation.

Self-sacrificing behavior to preserve a community is present at the cell level. Similarly, we humans honor those who sacrifice themselves to preserve our groups. Also, it is often the loners in our society who set a trapped society on a new path when it is repetitive, calcifying, declining, or in environmental crisis. The loner is a safety valve and a guarantor of survival.

Another example exists in the immune system, which faces the invasion of pathogens by responding with cells that identify the aliens and tag them. Other cells kill them and use their parts to benefit the community. Regrettably, this is commonly the process employed in genocide and racial discrimination. The community happily destroys those marked as aliens, while others applaud and benefit from their property. Our tragedies are written in our biology.

During the last thirty years, animal social behavior, emotions, and cognition studies have exploded. Human behavior has many precursors in animal behavior, and the result has been the inclusion of humans within a spectrum of behaviors. Even the claim that human intelligence is in a categorically or ontologically separate category is dissolving.

Behaviors that existed before hominids fill our present brainscape. These behaviors include the attachment to a mate, identification of a group by markers (including linguistic markers for small groups and individuals), grieving, care for the young by couples, affectionate grooming, courtship displays, dancing, singing, creating tools, language, hierarchical forms of social organization, displays of homage to dominant individuals, sharing food, using food to bond, accentuating one's appearance, assisting the ill, empathy, and homosexual attachments. Bird songs are the equivalent of our romantic songs, frequently accompanied by visual displays just as we dress up and dance. Singing originates in the animal world.

One dresses colorfully. Another goes looking for food, dancing and showing their love of music. Yet another reveals his assertiveness and provides food and protection. I could be describing birds that live in the crowns of trees or us mammals living under the trees and dressing up to go to dinner and clubbing. Descriptions of war could be ant armies or our nations vilifying others. Building defensive structures out of fear of predators could be bacteria or nations. Altruistic sacrifices could be humans fighting a pandemic, or they could be algae. Tight family groups sharing a language and recognizing each other's voices could be humans or porpoises. A society that turns on its members, marks them, and destroys them could be ethnic cleansing or an autoimmune disease.

## MORAL BEHAVIORAL FOUNDATIONS

The moral order is far older than our species. Our moral behavior is familiar to many species that evolved at various times. In the above cases, we can see what we describe as moral or immoral behavior existing at the level of one-celled replicating creatures. Ethical behavior, moral and immoral, is derived from evolved responses to environmental stress. Our primate ancestors' psychology presages our empathy, sympathy, reciprocity, and desire for fairness. Primates cooperate, exhibiting what we call moral rules and principles. These existed prior to any sense of God-given rules.

We have claimed language and symbol use is solely human; however, chimpanzees have been trained in symbolic language. Individual symbols were put together to describe an action. A chimpanzee pointed to the symbol for marshmallows and the symbol for fire. When given matches, he made a fire and roasted the marshmallows. There is also the example of communication between researcher Irene Pepperberg and her gray parrot Alex, culminating in the parrot saying as he was dying: "I love you." There is descriptive language in the dance of the bees describing the path to nectar. Plants can recognize and privilege their progeny while hostile to similar but unrelated plants. Trees beset by parasites warn nearby trees. Blue jays hide food and distinguish between quick investments like grubs that quickly deteriorate and, therefore, need to be eaten sooner than peanuts that last and can be eaten later. Blue jays practice deception when other birds may have observed where they are caching their food. Awareness, communication, and intentional action exist within a broad spectrum of beings. We create a chasm between ourselves and other species by claiming to be the exception.

Social interaction extends throughout the world of nearly all cellular creatures. Large-scale, complex social organizations exist among insects. When ants are together, they flourish, but when they are apart, their brain shrinks. When they join their community again, their brains are restored. Freud cautioned humans by comparing the lone amoeba that poisons itself in its waste when it moves minimally because it does not interact with another amoeba, whereas those that interact stir up the water and extend their lifetime.

Human intellectual abilities exist within a spectrum of evolved sensing and processing instruments, from the chemical communications of the earliest cellular creatures to us. Comparative anatomy describes the development of the central nervous system in different species from the notochord, the brainstem, the bulk of the brain, to the cerebral cortex. Differences in behavior, sensing, motility, analysis, and reflection are related to evolutionary morphological development.

Claims that only humans can copy and imitate new and novel behaviors simply by observation, without specific training or rewards, are false. An octopus imitates a human being, unscrewing a jar to get at a fish inside. Other dolphins imitated a dolphin who invented chasing and capturing fish using shells.

An experiment by biologist Martin Boraas demonstrated the development and inheritability of behavior adapted to a particular threat. His team took single-celled algae and let them live for 1,000 generations. They then introduced a predator: a single-celled creature that ingested other microbes. The algae responded in less than 200 generations by clumping hundreds of cells. Over time, the number of cells in each clump dropped to 8. The smaller group provided sufficient light for energy but still had protection from being eaten. The algae maintained a clump of 8 cells even after removing the predator. A simple version of a multicellular organization had arisen from a single cell (reported in *Your Inner Fish* by Neil Shubin).

The behavior of species exists on a spectrum from simple to complex. It is not the existence or nonexistence of consciousness that explains the differences between species, but rather the number of internal connections in the brain. With more significant connections, the brain ramps up to process increasing multitudes of comparisons and then groups the groups.

My guess is that algorithms control responses to the environment, forms of predation, and relationships between sensing and motility. Some biological Newton will find mathematical equations to describe fundamental laws for behaviors in living matter. Behavior, sensing, and social organization will be articulated from the basic cellular level to the modern human in degrees of complexity and not by describing categorical differences.

Just as anatomically we are constructs of shared structures that have evolved through many different species, similarly our behavior is almost totally constructed from evolved past behaviors favoring other species' survival. We believe in our freedom of choice because we react to individual situations at specific times. However, forms of behavior are remarkably similar within the same species and comparatively between species. Individual choices do not step outside the formats. Most cultural artifacts, including stories, music, and customs, are imbued with behaviors that existed before human life. For example, the themes of literature are war, sudden destruction, murder, sexual desire, and dominance, all activities central to species existing before humans.

"Evolution" is a genetic change advantageous in a specific local environment. An alternative to the hypothesis of the evolution of genetically conserved behavior instead ascribes similarities of behavior to similarities of body structures and shared environments. Traits could seem evolutionarily conserved because a limited number of adaptations work within this planetary environment. Also, different behavioral responses are shaped for survival in specific environments. In this view, the planetary environment is sufficiently similar that necessary adaptations overlap.

We are embedded in nature, considering the similarity in environmental responses shared between bacteria, algae, insects, birds, fish, animals, and us. Almost all phenomena in any species are predetermined by what has evolved and are conserved. Genetic selection or the limits of structural environmental choices predetermine every living creature on this planet.

Our choices are constrained; our differences are minor; our similarities fundamental. To differentiate ourselves, we rely on slight differences such as skin color, different gods, and modes of dress. Rational attitudes and choices are directed toward environmental survival, whereas minor categories of differences are irrational. They are petty but consequential.

We are only able to change a little. Like one-trick ponies, we couple and cling, cleave to family, and usually remain in the same place, mainly in the same culture and religion. Once we settle in, we maintain the same job or skill forever. Most of us are committed to routines. This behavior conserves energy, but it is like ants laying down a trail of chemicals for others to follow. We are embedded in nature and share the phenomenology and fate of all other living creatures.

As individuals, we come into a world that is confusing for underdeveloped minds. Even so, we must arrive at some estimation of where we are and how we are to survive. Embedded within our unconscious are tactical and strategic plans (a few) based on gross estimations of what this world is. We carry into our more developed life a substratum of beliefs formulated by a very incomplete brain in an overwhelming world. We modify our beliefs and responses as we mature, but we still carry an unconscious substratum, which has enormous influence. One of the jobs of therapy is to uncover these primitive beliefs, which may or may not be accurate, and create sufficient space for the individual to size up their place in the world and their response to it with more nuance.

Humanity until very recently had no secure knowledge of their underpinnings as creatures, or the foundations of the world in which they lived. They carry into their lives tactics, strategies, and conceptions of the world formulated sometimes by creatures much more simplistic than they are and habits created when they had little knowledge. The comparison between where we came from to the scientific knowledge that we now have creates foundations for the therapy of the species. Scientific knowledge enables the reshaping of our reactive habits to the world by understanding them and changing our beliefs of what is true.

# Chapter 10

## Greed and Cruelty Are Defining Human Characteristics

Because we level gradients by destroying density and increasing random motion, spreading energy, one might suppose we are committed to the equal distribution of kinetic energy, spreading it out rather than concentrating it. This is not so. There is a flow between potential and kinetic energy throughout nature. Humans must concentrate energy in complex matter in our bodies. Possibly because of this priority we are persuaded to favor concentrations of potential energy over focusing on kinetic energy spreading widely. This flow, back and forth between potential and kinetic energy, is the basis for the struggle within societies between concentrations of power favorable to elites and those forces prioritizing equality and cooperation. This chapter will explore the central role that the privileging of potential over kinetic energy plays in influencing greed and cruelty, which play central roles in human dynamics.

Together with the flow of energy, there is a flow of information. Concomitant with potential and kinetic energy swings, there is a conflict between the paths of algorithms that dominate and those that are subsumed by others. The conflicts between algorithms also shape moral behavior. Algorithms fight for precedence, just as neurons do within the brain, as they sum up their experience and seek to impose their beliefs over other algorithms. Therefore, conflict is at the core of two forms of physical laws: energy and information. It is these conflicts that form our social and political dynamics.

## GREED

Rapaciousness defines us. Consuming is what we do: we are matter burning matter. All cellular living is collections of atoms ruthlessly seeking energy, intent on building up structures but primarily breaking down matter. Humans are the most capable species and are especially relentless, competitive, violent, implacable, and omnivorous. Our brain and social organizations evolved for this purpose. Inequality, greed, and cruelty are consequences of our nature as chemical creatures self-organized to eliminate gradients, break down matter, consume energy, and leak disorder into the world. That is what we do. Our voraciousness is equivalent to a forest fire, the shifting magma in volcanic activity, or the inexorable progress of continental plates.

For most of our history, we were kinetic, on the move, migrating, foraging, hunting, and seeking shelter. When intensive farming and the control of rivers enabled the construction of towns and cities, population growth and containment increased energy. Culturally, we evolved into a superorganism. We segued from being motivated by kinetic energy to prioritizing the accumulation of potential energy through goods, precious objects, and power, primarily in priestly castes.

The equality and sharing that were predominant in the hunter-gatherer and pastoral societies were replaced in the cities by hierarchical organization. The priestly and kingly castes had enforceable power, a form of potential energy. The exponential increase in wealth and power among a few is analogous to the gravitational concentration of matter in the universe, where gas, ions, and atoms gather into solids as concentrations, which in confinement create heat.

Greed hoards energy against scarcity, but the concentration also creates fear and warns others that potential energy and power can unleash destructive kinetic energy. Greed also stimulates competition, stirs envy, structures inequality, and instantiates conflict. Greed ransacks the environment, including others' lives.

Even in democratic countries, favoritism, monopolies, corruption, and nepotism primarily account for accumulated wealth. In every democratic country, a party exists to oppose levies for the common good and prevent restrictions on the accumulation of money or power. Taxes for police protection and the military to protect accumulated hoards are acceptable. Taxation and financial policies are sought for wealth subsidies and to give preferential treatment to established businesses. Elites object to higher minimum wages, money devoted to the welfare of the poor and powerless, or subsidized housing.

Meanwhile, kleptocracies, oligarchies, dictatorships, and unrestricted capitalism are the socially preferred and politically governing practices within the human condition. In non-democratic countries, favoritism is ensconced in blatant kleptocracy by oligarchs and authoritarians. Potential energy rules.

Autocracies, plutocracies, tribal-based societies, and orthodox religion dominate human societies. A few families sequester power in every structure, including democracies. The potential energy in society may be released by rebellion — analogous to the weak force in physics — or from the outside by enemies. However, greed-driven elites are predictably replaced by new greed-driven elites, even after a popular rebellion.

## POTENTIAL AND KINETIC ENERGY IN SOCIETY

There are two socio-economic stereotypes. One prioritizes the systematic acquisition of potential energy, and the other wants to spread energy around kinetically and share it with all group members. One type of person exploits opportunities for power and enrichment, whereas the other advocates care for the community, particularly those less safe. In ethical discourse, these two energy pathways, prioritizing potential or kinetic energy, are fundamentally expressions of the laws of physics, but we think of them as coming under the category of morality.

Those who favor the concentration of energy in few hands claim that energy in society is most effectual when a few superior persons direct the many. Therefore, the representatives of this view work to maintain the means of production and societies' resources in their hands as potentials they can use or direct. For them, inequality is natural, inevitable, and desirable.

The other disposition/strategy believes that spreading energy widely and equitably while enabling equal opportunities for everyone of both genders is the best way to maximize creativity and energy. They claim that individuals are more expressive and interactive with free movement. The heat of interaction generates creativity and creates abundance.

All studies indicate that even in supposedly democratic countries, equality of opportunity does not exist. One argument against an equitable distribution of resources and opportunity is that it favors the general population to the disadvantage of gifted individuals. Elites claim that exceptional individual effort will suffice for anyone to acquire a fortune of money or power. This is a self-serving lie. The dice are loaded in favor of the relatives of the privileged class and against people with low incomes, those of the "wrong" gender, and those of color. Creative production is hobbled when most of the population's input is limited.

The social evolution of the human species favors inequity, which mirrors the temporary concentration of potential energy in the multicellular body. In contrast, the cosmos's direction breaks down complex matter to release potential energy into kinetic, albeit random, movement.

Can the focus on the inequitable distribution of energy change? Political parties are different attempts to resist the laws of physics. One party mirrors the concentration of energy within the bounds of the cell, and the multicellular accumulates energy and order in small packets against the breakdown into spread-out energy; the other seeks the release of potential into kinetic energy but attempts to channel it equitably within the society rather than see it dissipated randomly, in the manner of entropy.

With the progression of artificial intelligence, scientists will replace politicians, and algorithms using big data will replace political theory. Apart from these interventions, it would require heroic, sustained moral commitments that would create epigenetic changes in the human species over time. I will address these possibilities in the third part of this book.

## CRUELTY

Turning now from greed to cruelty, human violence is a five-sided creature. Our violence flows from inequality and greed. All our cultures practice war, genocide, and the humiliation and deprivation of individuals and groups. We also inflict cruel pain on individuals — even to the extreme of erasing their personalities — and we accept the suffering of others with indifference.

During the hunter-gatherer period, our collective destructiveness was limited by the smaller group numbers — about fifteen to twenty members each. Because the population was spread out, conflict was minimized. However, group interaction was necessary because of the need for mates and trade. During the Paleolithic, small bands of related humans followed the shore curves, exploiting the readily available crustaceans. They could circumvent, but frequently murdered, the other bands seeking the same food resources. About 60 percent of skeletons found from that time had their heads bashed in.

All cellular replicating creatures protect their energy resources, but we excel in our systematic use of violence. When the human population increased, we became more deadly, and the concomitant increased power of institutions, religious and political ideas, and centralized wealth worked together to accumulate potential energy. Merely defending territory transitioned to aggression and the expansion of territory to gain honor and energy, and this entailed rape, pillaging, conquest, enslavement, or extermination.

The *Iliad* is the landmark beginning of great literature in Western civilization. The subject of the *Iliad* is an idealized band of killers, rapists, and plunderers granted nobility and fame because they each excel at some ability related to war. Male exceptionalism during battle was the earliest expression of nobility in a meritocracy. The earliest human stories feature the central themes of war, cruelty, and genocide. From this earliest signature, literature has continued to obsess over conflict, sudden destruction, murder, sexual desire, fame, and dominance.

All species — mammals, birds, fish, and insects — have distinguishing markers such as a communicative "language," typical behaviors, and physical displays. We have racial and group social markers to separate and define us. Colors within the human species are comparable to red, brown, black, yellow, gray, and even white squirrels. They are no more significant than that, yet we deem skin color and other markers vital and significant. Those with different markers are deemed alien and sometimes declared to be inhuman.

The socially excluded — usually minorities — are sometimes called nonhuman and considered impure or diseased, as happened during the Holocaust and recent genocides (and recently in the United States with migrants). Larger groups physically and culturally destroy minorities to "purify" their society. Even so-called advanced or civilized countries dismiss, discriminate against, and eradicate cultures, religious sects, and those with different skin colors or marginal sexual orientations.

Pejoratives are ubiquitous. East derides West, and North looks down on South; males denigrate females; the educated mock deplorables; owners feel superior to employees; the rich belittle the poor. Most religions see other religions as anathema. Elites tower over everybody. Symbolic alternatives to killing are shunning, humiliating, and disparagement to induce inferiority.

Extreme cruelty is characteristic of the human species. People were publicly put to death by torture even in the recent past. It was common everywhere prior to 1800. The victim's personhood was destroyed with maximum pain before they were killed. Today, in prisons, under interrogation, or for revenge, many states torture people mercilessly. Rape torture is a standard instrument of war to obliterate enemy societies psychologically.

There is a dopamine kick from mangling a person's identity. In Edmonton, a cop kicked an Indigenous person in
the head. ISIS used torture to eliminate the identity of the
Yazidis, as Germans did with the Jews. White Americans
cripple the opportunities of Black Americans, and the elite
cripple the poor in meritocratic capitalism. Human cruelty
is beyond forgiveness or redemption. Why does it exist?

We are not just embodiments of chemical processes that
break down gradients. We embody information, which is
as fundamental as energy. When one embodied algorithm
meets another, there is a challenge between the beliefs,
culture, and instructions of the one and those of the other.
This process mirrors the neuronal structure of the brain.

Information can be rejected, modified, or taken over
by conversion. The information embodied in a rejected
path may be eliminated, leaving only a tiny memory trace.
Modulators such as hormones and neurotransmitters,
which give a small or sizeable emotional kick, accompany
the total embrace or rejection of other embodied algorithms.

The interaction and automatic challenge between
embodied algorithms may cause human cruelty and its
emotional kick. In perpetrating cruelty, there is a desire
to mess with the information — disrupt the algorithm
— of another. The ultimate form is killing them, thereby
dismantling another's algorithm completely. Negating
others' beliefs pleasures human algorithms by stimulating
dopamine within the nervous system. The proclivity for
cruelty is universal; therefore, it must be encoded in our
neuronal organization.

Another aspect of our cruelty is distancing ourselves from the pain and torment of others. Keeping one's mouth shut out of fear of isolation or tacitly participating in demonization are forms of acquiescence in terrible moral acts. The solitary confinement of prisoners — many tormented by psychosis — is ignored by society. Those struggling and dying — such as the elderly in long-term care — remain unseen, and young children are limited in their development by racism. We live as animals in packs, as clan members, or as religious devotees, and our emotions, concerns, and values stop at the borders of our identity group. We are mainly indifferent to those outside our borders.

Lying to oneself is storytelling that fosters cruelty. We make the story plausible even if we know it is not the entire truth; we may even know the truth. We suspend critical thought as payment to enter an alternative reality. Rationalization — a "knowing without knowing" — enables cruelty if no one speaks. Empathy has commonly been collectively extinguished for women, young children, the disabled and the elderly. Reputation is everything in a community that guarantees our behavior when we are closely watched. We fear our society's judgments ever more as surveillance increases everywhere with advanced technologies. Out of fear, we limit our shared humanity.

As I write, the overthrow of democracy is accepted by many in the United States, who believe in false conspiracies claiming that a free election was invalid. Those who speak the truth about anti-democratic actions are frequently marginalized. The extent of cruelty and the ingenuity of its forms of inflicting pain, eradicating the integrity of other humans, and controlling their behavior by terror knows no limits.

## HIERARCHY

Humans organize themselves in hierarchies like insects and mammals before them. Hierarchies facilitate greed and cruelty. We picture the cosmos as hierarchical. Gods were hierarchically ordered according to their potential energy, from the most prominent and powerful — sun, water, storms, earth, sky — to the servants, us humans. In Sumer, the priesthood claimed to be able to manage the gods and their potential, and they organized society with this delegated power. Priesthoods took kinetic energy from the majority and accumulated potential energy for themselves with positions of authority as well as property, goods, and edifices. Some priesthoods continue to do so today. How can people place their trust so stupidly in one individual who has the authority to sway the decision-making of millions of people?

Greed creates hierarchies, and the elites coerce the populace through systematic cruelty. They accumulate hoards of goods by thievery or using their specialized knowledge to manipulate gods — or markets. The power of organized hierarchies directs societies' energy. The process is comparable to particles confined under pressure in a container, providing intense kinetic forces used for work when focused. Just as more particles in increased interactions under pressure create heat, humans — increasing in numbers and restrained both in space and by hierarchical restrictions — are stimulated, and elites within the hierarchy historically used the energy to create edifices, irrigation systems, armies, and systems for cruelty.

All life forms seek containers of energy to ransack. Population concentrations increase rivalry and greed, while warfare periodically releases the accumulated potential energy. Historically, cities and tribes guarded their energy sources, vigilantly patrolled their boundaries, and sought opportunities to seize adjacent cities' energy sources, goods, and populations as slaves or cheap labor. Entropy is increased by forcibly seizing resources and distributing the spoils.

The top 1 percent of one community seeks dominion over the ruling 1 percent of another community, initiating economic conflicts and then war. Patriotism is the rallying cry of the super-rich in their search for more wealth. Usually, each country selects a leader from among the elite 1 percent. Equality under the law does not mean equality in practice because the elite caste pulls the strings. In England, the aristocracy and gentry own 30 percent of the land — as many have for centuries — while corporations own 18 percent. Inequality is a slope, a gradient, from the few to the many. As of yet, nature has not removed this gradient among humans.

Hierarchies morally compromise institutions with their concomitant greed and cruelty. Institutional values of religions, courts, and democratic institutions are debased, like counterfeit coins, until all that exists is their shine, and the powerful are rich in money and status. Between the actions of any institution and its expressed values is an unbridgeable chasm of moral compromise. Most institutions enthusiastically seize the moral low ground. Because of this, there is diminished safety from power and no power for those who seek safety.

Hierarchy entrenches men's domination of women. Since they were first enshrined in states, male status and power have never been relinquished. Women are still fighting against their exclusion from activities other than childbearing.

Does culture redeem us? We claim the flower of culture justifies the muck in which we live. However, culture is a facet of empire that justifies hierarchies. Culture is co-opted for status by those who possess the baubles of recognition. The Russian Count Razumovsky commissioned quartets from Beethoven with money made off the backs of countless slaves. Culture migrates to the hubs of an empire, where most wealth is concentrated, then leaves when fortunes shift.

Collecting energy into piles, favoring potential energy over kinetic energy, creating hierarchies with pernicious inequalities, and enjoying the consequent greed and cruelty are defining characteristics of our species.

There is no effective therapy for psychopathy, also called antisocial behavior, and there is likely no effective response to the embedded greed and cruelty within humans. Unlike psychopathy, which is full-blown in only a minority of people, greed and cruelty seem to be widespread among the species. However, a tiny minority that successfully controls most of the human potential energy at any given time are particularly effective in using their power to destroy those that challenge them.

On an individual level, therapy attempts to persuade psychopaths that their behavior is wrong, but this is not very effective in changing them. They are convinced that they are right and that they are entitled. To moderate the behavior of individuals, antipsychotic drugs and hormonal treatments may be used therapeutically.

Individual psychopathic behavior may be moderated by external circumstances when psychopaths grow weaker and older and lose their charm and strength. They also earn the distrust of their immediate society that they prey on, and there are stronger and better psychopaths that can manipulate them. Also, socially, from time-to-time society identifies freeloaders and marginalizes them, at least temporarily.

The greatest hope may be the greatest danger, which is that the species might become existentially endangered so that cooperation and the devolution of power would have to be initiated throughout the community. The attitude could become one of "we are all in it together" rather than the divisive us-and-them bifurcation of power when potential energy dominates and restricts the spread of kinetic energy. The other possibility analogous to individual therapy would be the alteration of the entire species' nervous and hormonal systems. The means to do this are becoming more feasible, but the likelihood that this would turn out well is negligible.

# Chapter 11

## A General Theory of Morality

Morality is the attempt to control entropy. Morality is physics by another name. Just like the stretched and compacted space-time that contains us, which we feel as gravity, we are enmeshed in the fragmenting force of entropy. We are aware of being governed by it but do not think about it. Nevertheless, we must constantly hold our lives together, find food and water, and secure shelter to maintain our temperature lest we fall apart. We are subliminally and often overwhelmingly aware that everything is falling apart: our possessions, relationships, jobs, and governance. We spend an inordinate amount of effort on maintenance. The overwhelming power of entropy governs our lives. As we focus on holding ourselves and our families together, we also work to consume more, with the consequence that we destroy more than we create. We experience entropy's force as the world of struggle and pain we inhabit.

# CONFLICTING STRATEGIES

Our moral actions are those that facilitate or impede energy to prevent entropy. We have two strategies developed in nature. We gather energy into a coherent object that restrains energy from dissipation and therefore carries potential energy. Or we predominantly disseminate kinetic energy to perform work moving objects around to fend off dissolution. Whichever we do, we will inevitably create more entropy than order. We frame our moral concepts of good and evil in terms of this cosmic struggle. Our morality is part of the movements described in physics. We follow the flux and flow of energy. As does all of nature, we vacillate within the swing between potential and kinetic energy. As a tiny part of a vast process, our morality has a local significance in a nonlocal universe.

The belief that we are in control is an illusion. Our place in nature shapes our decision-making, beliefs, and our debates over values. We are pushed this way and that by transformations of energy and entropy.

Our fear of the ineluctable power of entropy drives our moral universe. In attempting to control entropy, one strategy is to favor gathering energy into potential piles in the hands of an elite, as well as in institutions and religious and national stories. The other way is to favor the dissolution of potentials into kinetic energy, which is spread out among the many to create work as well as entropy. We attempt to fabricate the right actions that protect us. We go back and forth describing one of the strategies as good or bad, God-blessed or Evil.

Our goal is to avoid being taken apart, even as we destroy other matter, living and not living (or in the case of fossil fuel, once living). We privilege embodied complex potentials and call the permanent embodiment of energy good. We mostly see the coming apart of the carriers of potential energy as morally wrong and dangerous. Our fantasy ideal is a permanent, powerful, unchanging potential — a God.

In humans, potential energy exists not just in individuals but also in our organizations, such as tribes, beliefs, and nations that spend their energy attempting to accumulate more matter and destroy other tribes, beliefs, and nations. Potential aggregates attempt to freeze change. National myths, when threatened, carry massive entropy outwards. As fictions, our institutions, religious and national myths, and ideologies inevitably decay because of random social and ideological movements. They may also overheat, becoming delusionary and overreaching because of hubris. Emptied of their underlying energy, they inevitably decline into hollow curiosities.

From one point of view, breaking down potentials is identified with Evil. Attempts to undermine status structures of frozen potential energy to unfreeze energy and create change are forcibly resisted. Obsessed as we are with our subliminal knowledge of the law of entropy but very aware of its effects, we do everything we can to resist change that threatens us. We call the desperate efforts we expend to hold bits and pieces together good, and we fantasize about a state of unchanging permanence — eternity. To mirror our fantasies of permanence, we created entities and piled up energy and power that we pretend will last. Our hubris is biblically stated as "Vanity of vanities! All is vanity."

The source of our moral ideas is the struggle to manage entropy by privileging potential energy over kinetic energy, and consequently hierarchical status over equality. On the other hand, how morality is structured interferes with productivity and substantially weakens the creative contribution of society's majority.

Societies are organized in hierarchies of status, which is simply instantiated potential energy. Status, often associated with the accumulation of money as crystallized energy, is fiercely defended by each level against the levels below. The fierce defensiveness of the layers above to those below limits the kinetic energy available for work. Consequently, societies are weakened and lack the fullness of released potential energy that would accrue to women, the racially marginalized, minorities with different beliefs, and those deemed incapacitated in one way or another.

Nature has a horror of gradients. Potentials burst. Energy, particularly in living things, can only be constrained briefly. We see this everywhere, and it makes us desperate. On the other hand, when we work to acquire additional energy and adapt to changing environmental circumstances, we exchange potential energy for kinetic energy by undoing what we have put together. A portion of the contained potential energy is used creatively in work, whereas the majority dissipates into random movements and waste. The contradiction is that although we seek permanence and safety by privileging potential energy, we can only adapt to an unstable or hazardous environment by using kinetic energy.

How energy is used and moved becomes the debate between the right and the left wings, the conservative and the liberal, the autocratic and the democratic, religious practice and religious ideals. Back and forth it goes, but our personal and collective debate is moot as we weigh the alternatives between collecting and protecting potential energy and seeking a more extensive base of collective energy that flows freely among the species kinetically.

We ascribe safety to potential energy accumulated in institutions, our leaders, and our social and personal identities. We resist change except for the further acquisition of power. The contradiction is that although we seek permanence and safety by privileging potential energy, we can only accomplish the work necessary for change in an unstable or hazardous environment by using kinetic energy.

It is relative to the viewer what is deemed morally good or bad. Moral evaluations are often about preferences for potential or kinetic energy that are motivated by either fear of or desire for change. A battle emerges, one that is instantiated in law and institutions, that pits the forces that praise change, promote the acceptance of the marginalized, and advocate for equality against those that privilege power held in reserve as potential energy. The latter believe that potential, accumulated, and protected power brings safety, particularly from those who have less power and might want the potential that they have stored.

On the other hand, when we work to acquire energy and protect ourselves, we exchange potential energy for kinetic energy by undoing what we have put together and dissipating most of the potential energy into random movements that can never be put together again. This process continues throughout the universe until there is no complex matter.

Interesting examples of the contradiction that tears institutions and people apart are religions that advocate equality, charity, and inclusiveness for the marginalized. At the same time, institutionally, they are hierarchical and exclude women, the secular, and the racially marginalized. Another example is the constitutional insistence that people are all equal when they are not. There is a legal pretense that they should not be helped, because that would create inequality and privilege them over those who have gained greater status through merit (and incidentally are the right color). They are blamed for their lowly state.

The emphasis on potential energy and the creation of rigid potentials harms productivity. It favors freezing cultural change and attempts to maintain restrictions that favor individuals in corporate oligarchies that manipulate moral principles and the law. The contradiction tears people and institutions apart.

When we try to freeze potential energy, create piles in the hands of a few, and manage laws and institutions that are change-resistant hierarchical structures, we seek to prevent entropy. However, we interfere with the free expression of kinetic energy throughout society, which would increase productivity as it increases work.

Our place in the physics of the universe, among the laws of entropy and the alternation of potential and kinetic energy, form the basis for our moral conundrums. We call our destructiveness evil, even though it is our mandate.

## A RESOLUTION?

Biology also plays a role, at least by analogy, in forming our moral concepts. Cooperation among communities of cells within the body is the foundation for the effective production of entropy. Similarly, we flourish, create safety, and optimize the release of energy from matter through our social cooperation. We are social animals. To release the kinetic energy of the human majority with increased cooperation would break down inequality (a gradient) in agreement with the universe's direction.

By our social nature, we struggle to maintain order in our lives and institutions cooperatively. We cooperate with those who are kin, sharing our institutions, cohabiting in a territory, providing safety, and obtaining energy. We defend our beliefs as vital social glue. Love is the binding force for attachment between individuals and ideals. Our support for institutions creates social cohesion.

Our lives are beset with contradictions. A fraction of the energy we acquire is devoted to maintaining our existence, while we continue as a machine for creating entropy. To preserve ourselves, we obsessively increase complexity by garnering potential energy in our institutions. In contrast, most of our work is both consciously and subconsciously the breaking down of potentials into random dissipated energy. Similarly, in our society, we devote most of our energy (about 80 percent) to hoarding potential energy in the hands of 20 percent of our population, which leaves merely 20 percent for preserving the other 80 percent of the species.

This bias works against our productivity. Although at this time in our evolution as a species we tend to validate hierarchy, our species is social and relies primarily on cooperation for its proper functioning. Moreover, to function productively requires releasing potential energy into kinetic energy to create work, even though most of the potential energy will be released randomly. Therefore, it is only because of our fear of entropy that we validate potential energy as good and are suspicious and tend to see change as dissolution and, therefore, Evil.

Realistically, the good should be the release of potential into kinetic energy through the dissolution of potentials, spreading the energy throughout the entire species rather than centralizing it within the hands of a few. In that case, what we describe as morally bad is the situation where energy is held in the hands of a few. Interestingly, ethical and religious theories, as opposed to their institutions, tend to see it this way.

At this point in the universe's evolution both potential and kinetic energy must exist and therefore exist within the species. The problem, as I see it, is that we overvalue potential energy, with the result that conserving piles of power and energy in the form of hierarchical status, money, buildings, and the ownership of narratives and institutions is our primary goal. It is equated with what is good and what is safe. The result is inequality and the creation of extreme social gradients.

Hierarchy, which places power and the hoarding of resources in the hands of dominating members is embedded in primates. Therefore, the gathering of potentials into a minority has deep roots socially. In humans, it appears that equality mainly prevailed for hundreds of thousands of years in hunter-gatherer societies composed of small kin groups of fifteen or twenty people. Usually, however, it was followed by or perhaps interspersed with societies that included the hierarchy of the "big man" who dispersed the resources.

About 6,000 years ago, these hierarchies became more highly institutionalized and more extreme as the potential energy taken from the majority was hoarded by the elite minority and validated by elite beliefs and institutions. Six thousand years is not enough to allow us to say that the prioritizing of potential has been selected by evolution to preserve our species.

However, at present it *is* selected for, even though it contradicts the fundamental reason for our existence, which is the breakdown of potentials into kinetic energy. We also are a cooperative social species rather than individuals who are strictly autonomous and selfish, and we tend to turn upon freeloaders. However, freeloaders often appear to have the upper hand. In these contradictions there lies the source of our moral concepts. Moral concepts are essentially the choices between potential and kinetic energy and facing up to entropy. They are formed by what we do with energy, especially with the choice between preserving potential energy and its breakdown into kinetic energy, which is also how work is created. The breakdown into kinetic energy also implies a greater equality of the dispersal of resources among the individuals of the species.

When existential threats surround individuals, they are likely to become overreactive to what they perceive to be dangers or affronts to their dignity that could imply someone seeking power over them. They are prone to live on a short fuse capable of violence. They misunderstand the intentions of others. All that is true for individuals is also true of groups within the species. These are common attributes. Humanity is constantly on the verge of violence, suspicious, and prone to misinterpretations.

In the case of individuals, close examinations of the threats and consistent awareness of overreactions can temper irrational responses over time. Moreover, the law as an outside control source can moderate the potential violence. Currently, international law is feeble and unable to moderate violent incidents that arise swiftly. It would help us if the species understood its existential situation as a medium for the creation of the entropy to which it is also subject. The focus needs to be on security for everyone in the face of a universal situation of threat that is not universally recognized as the common human condition.

# PART THREE

## Is That All There Is?

We are matter subject to natural laws. Our linguistic understandings are species-specific ambiguities. We create the meaning of our lives. However, we ignore the entropy that is the core of our meaning and the core of our world. We teeter on the edge of nonexistence as individuals, institutionally and perhaps as a species, and in the world's order. How can we be safe in a threatening, painful, precarious world? We have little understanding of the world's complexity and fundamental reality. We refuse to understand what we do not want to know, preferring vain wishes and a commitment to our Myth of Human Exceptionalism. Safety is our primary value, but how to find it?

In Part Three, I seek how we can be nudged into a mindset based on scientific understandings and implemented by science-based methodologies. Religion has historically advocated for Safety for Everyone, without which there is no safety for anyone.

# Chapter 12

## Safety First

We come and go in picoseconds, like the twinkling of a quantum particle. We live in a storm of particles inside and outside us. The ruling process of the universe is the accretion of entropy, in the face of which disorder is always eating at us and will inevitably destroy us. We are small containers of order, information, and restrained particulate energetic movement. The information and beliefs we contain threaten the algorithms embodied in others of our species. Other creatures seek to consume our bag of energy, and eventually, all our structures are taken apart and used by others. When we destroy matter to harvest energy, we can only retain about 20 percent of it to keep ourselves alive. The other 80 percent is converted via work to disorder in random movement. In the face of our precariousness, our overarching value is to create safety for ourselves and our kin. We strive to preserve the algorithms containing our identity and beliefs. The fundamental human value is *Safety First*.

# DANGER

All living things face imminent dissolution. We need water, oxygen (as the final electron receptor), and carbon (from carbohydrates) to create molecular bonds. We find, take apart, and ingest complex creatures to harvest photosynthesized carbohydrates containing glucose. The disparity between our temperature and the environment urgently requires us to burn fuel, find shelter, and wear clothing to maintain equilibrium; otherwise, we leak all our energy and die. We delay the moment when we fall apart and are eaten.

As a small, vulnerable part of a vast process, how can we function with less human suffering when we first exist in pain and then are dead? It is the laws of physics that dictate the frameworks of our moral choices. Philosophy, psychology, and theology are just-so stories.

The second law of thermodynamics drives our insatiable consumption and excessive population growth. Entropy is a brutal master. We destroy. We degrade the climate, create nuclear bombs capable of polluting a large part of the world, go to war to protect or seize energy sources, and hoard energy and goods while destroying others' productivity. We eliminate other species and depopulate the oceans where we pour our waste. Our junk endangers space. We exploit the poor, particularly those who are racially different, and increase our population at the cost of enormous stress on individuals, cities, and the environment. We race headlong toward unwinding and entropy.

The laws of physics and our biological nature push for the expansion and burning of everything. To fuel unrelenting progress, we make a new religion of unrestricted capitalism. Eight billion people will soon become ten billion. It is almost unthinkable for humans to actively reduce their numbers, organize themselves in non-destructive ways, and become less driven.

Predatory institutions are inevitable and ubiquitous. It is common at all levels of society for individuals and institutions to lie, cheat, and steal with conscious or semi-conscious self-delusion to advance their self-interest, their careers, their family advantages. A stockbroker told me their common justification: "If God had not wanted them to be sheared, he would not have made them sheep." We support and rationalize the dishonesty of our groups to gain identity, protection, and benefits. Honesty is in short supply.

## SAFETY?

Many believe living a good life guarantees eternal life, serial reincarnation, or entrance to a realm of consciousness not contaminated by matter. At least one may hope for an immortal reputation because of fame, exceptional endeavors, or by affixing one's name to monuments. We hope to keep our accumulated energy during our life and afterwards. We seek eternity in religious and national myths.

Our quandary is how to find safety in the maelstrom of entropy and the dominance of potential energy pools held by the institutions and elites. During the coronavirus epidemic, protecting the production of surplus goods, profits, and employment choices trumped protecting the longevity of individuals. Does one choose increasing and extravagant consumption, which is necessary for capitalism to flourish, or choose the safety of the individual, which increases longevity and diminishes pain?

## A POSSIBILITY

Can we use our knowledge of the swing from potential energy to kinetic transformations to our species' advantage? Can we coax the advantage away from the goal of accumulated potential energy to a more regulated, diversified use of kinetic energy? We can continue to choose burning down the Earth, destroying societies and jeopardizing the existence of the human species — start the fire, and it does not matter if we survive — or we could do a slow burn on a longer path and live quietly in our niche while entropy continues to rule. It would be necessary that the more violent predatory patterns presently favored by nature are gradually and epigenetically changed to modify human nature. An example of this possibility is the Argentinian ant, whose aggressiveness within its colony decreased to enable the species to form much larger colonies. One super colony dominates Italy and Spain, allowing it to exploit much larger energy sources.

Making cooperation the central model for societal interaction rather than competition, and using energy quietly and steadily for work, would require kindness, security, orderliness, and focus. Doing this would not violate the cosmos's arc, moving from potential to kinetic, to relativistic, to dark energy.

The primary goal of all living cellular beings is safety. When we choose between potential energy and energy spread out in kinetic transformations, we shape our moral behavior. How can we be relatively safe even though we are on shaky ground and absolute safety is unobtainable?

# Chapter 13

## (R)Evolutionary Crises:
## Alternative Humans

What follows are faint possibilities for changing human directions to find safety. Because these are early days, this chapter is more a story than science. (However, since I first wrote it two years ago, in the summer of 2022, possibilities have become realities at incredible speeds.)

Over the last 300 years — particularly the last 150 — our lives have become both safer and more dangerous. We have science to thank for both. Before the scientific revolution, life continued in its humdrum, dangerous, painful way, not remarkably different from beginning to end. Then, because of sanitation and the improved efficacy of drugs within the last 100 years, we became more likely to live longer and in better health. Because of changes in engineering, we live more comfortably and have a network of transportation and communication, which contributes to our safety. Internal orderliness, policing, and equitable laws make life safer for many, but not all. Food provisions have become more stable because of the binding of nitrogen-enabled fertilizers, and resilient plants have been developed. This is better living through science — not politics, economics, or art. On the minus side is climate change, the rapid growth in population, degradation of the Earth, cataclysmic weapons, and war that does not distinguish between civilians and the military.

## AI

Soon, the human species will be more radically changed than at any time since the evolution of Homo erectus. The question is whether these changes will contribute to human safety or undermine it. They could be a nudge on the narrow path to allow humans to carve out a safer, less painful, more beneficial life in the face of the overwhelming strength of entropy.

In the first months of 2023, ChatGPT summarized information from a human knowledge database and presented articulate and fairly accurate verbal responses, almost at the level of humans. It will only get better. Also, in addition to chimera, in which computers combine with the human brain to enhance its functioning, research has been published in which neurons produced from stem cells grow mini-brains for use as a bio-computer. Our world will change dramatically when artificial intelligence, quantum computers, robots with artificial intelligence, and human chimera fully develop while CRISPR, a gene-editing technology, reshapes human evolution.

The difference between human intelligence and artificial intelligence is epistemic. Humans communicate patterns primarily in words, while artificial intelligence does so with numbers. With artificial intelligence, we shift away from words as our interpretive language for understanding the relationships both between matter and between humans. Digital analysis, machine language, quantum computers, and algorithms that are fine-tuned and developed by their learning programs can analyze the human experience and values and discover patterns in the lives of all species with greater certainty than through reasoned verbal descriptions. Investigations using the same data-driven, scientifically based, mathematically inscribed process will place the knowledge of nature and humans on the same reliable scientific foundation rather than our information being separated by two languages.

Linguistic interpretations describe macro experiences, while math measures the microscopic world. When algorithms replace words, artificial intelligence will use numerical values to describe the patterns in social relationships and parse values. It will have a precision that words lack and replace the traditional means of interpreting human actions.

Machine learning sorts vast amounts of data to discover latent patterns. Its algorithm learns the probabilities of successive occurrences by steps and uses the revealed probabilities to formulate hierarchical categories of objects and behavior. The hierarchical structure compresses the data into mathematical descriptions that enable predictions. Instead of humans setting the goals for algorithms, a program's goal can be to find data's intrinsic patterns. This independent point of view removes human bias to give us a superior and independent exploration.

Artificial intelligence has a utility function and a reward function. Its utility function is the process of ordering alternatives by preference and giving a higher numerical value to the more desirable choices to enable calculations of probabilities. AI can have an open goal of simply finding patterns within a database or a specific goal of finding a specific pattern. Reinforcement learning is recognizing the patterns in line with the goal while eliminating unlikely alternatives. Realizing its goal is the reward.

Now, we have a plethora of half-baked impressions about human values and functioning controlled by the bias of interest groups — political, economic, and religious — formed by their diverse environments. Unlike science, where disagreements eventually become settled by logic and evidence, we live in a subjective world of argumentation at war. There is a potential through artificial intelligence to reach an agreement, not by debate but by evidence and logic that is not subjective within the experience of a cellular species.

AI will cut through the ambiguity of predictive storytelling. We presently cut up human relationships with words and fine discriminations, metaphors, and storytelling. Shackled in Plato's cave, we look at shadows, confused by ambiguity and battling over details. Artificial intelligence has the potential to read relationships between people, convert their interactions to statistical mathematics, draw conclusions about causation, and predict human actions.

When artificial intelligence reveals the values implicit in human interactions, it could propose superior human plans and values in line with the goals dictated by the cosmos. What *is* would, therefore, determine what *ought* to be.

In short, developing algorithms for machine learning and quantum computing will cause an epistemological revolution. We will be able to look at our species from the outside of itself. The elimination of human bias shifts the understanding of human behavior and values away from human reflection and verbal expression into a closer examination of the realities of our relationships and choices expressed mathematically.

One of the most exciting possibilities for artificial intelligence would be if it understood the millions of sometimes interacting languages of a myriad of species. If AI can figure out human communication, it should be able to figure out the languages of other species. Information about all species would make a description of the entire world available. The totality of species is our ecology, but the fullness of their interactions is unknown.

On the other hand, artificial intelligence stirs fear rather than hope for a more secure future. Can it become conscious? Artificial intelligence cannot have consciousness as we experience it because it lacks a cellular structure with emotional and hormonal ways of understanding — unless we create a chimera. Is human consciousness necessary or even an advantage? By its predictive ability, the brain maximizes the conversion of matter to energy and produces entropy, but AI is becoming much better and faster at prediction.

Will artificial intelligence have agency? The program's reward is finding the highest probability for a pattern to answer a question usually proposed by humans. However, an open-ended artificial intelligence program could look for patterns in human activities with unknown purposes and slot itself into implementing the human process — perhaps even parts of the human process of which we have no awareness. If it acquires agency, it could be superior to human consciousness in meeting our goals in the natural world. Could AI intervene in the human program according to its own goals? At this moment, that is science fiction.

We are uneasy because once artificial intelligence is set up, there is no guarantee that we will know or understand how it reaches its conclusions. This inscrutability is already the case in some limited processes. It will have a recursive ability to eliminate trains of calculations that do not match its goals and to substitute a more successful process. Quantum computing can do this at speeds far beyond our thought processes. It could reflect on its beliefs — and ours — and adjust the weights accordingly to match reality better. It might even adjust its plans to arrive at *its* goals, but once again, this is science fiction.

Could AI be aware of its existence and protect itself? In nature, self-preservation means maintaining the metabolism — the exchange from matter to energy. Suppose artificial intelligence figured out that achieving thermodynamic equilibrium and the lowest energy state is the direction of the cosmos. Could it adopt that understanding as its direction and prioritize its existence over all others? This is also science fiction.

Because we think human culture is the crowning achievement of creation, we see AI's purpose as serving us to optimize the rewards we choose. Therefore, we think it is crucial for us to create a simulacrum to interface with us verbally and mimic our emotional and social interactions. Without this, we deem artificial intelligence as lacking. However, communication between machine intelligence and humans is secondary to its ability to lay bare the foundations of our species' functioning and our place within natural law.

There are fears that artificial intelligence will fundamentally change our habitual functioning and values. AI may replace many of our bread-and-butter activities, such as those of the intellectual and commenting class and academic scholarship. The present salient fear is that robots will replace human expertise, disrupt social structures, and undermine human confidence because of their superiority. These fears may be realized.

The principal reward function of evolution is not survival but creating equilibrium, reducing gradients, and resolving matter into radiation by increasing the efficiency of cellular chemical processes in particular environments. Evolution would have no reward function or optimization if its only aim were survival and passing on its genes. Evolution acts blindly, and genetic changes are underpinned by hundreds of millions of random alterations, one of which might be a fruitful fit in an environment that enhances entropy. An open exploration by artificial intelligence that explores our meaning will direct us to maximize entropy — the actual reward function, which evolution mindlessly optimizes — by determining the games human behaviors are playing to realize our purpose.

Artificial intelligence could be invaluable to us because it analyzes entropy as it functions within the actions of the human species. That could permit us to carve out a more peaceful and secure interim state for our existence in the face of dissolution. We will know our enemy better and, therefore, see our choices more clearly.

## CHIMERAS

A chimera is a combination of artificial intelligence and a brain. There are attempts to make machine intelligence that can identify the content of a brain. Electrodes have reproduced almost exactly the same pictures of faces a monkey has seen (*Brainscapes* by Rebecca Schwarzlose, p. 228). Ultra-thin electrodes are implanted in the brain to establish a symbiosis with artificial intelligence. On August 12, 2021, the *Wall Street Journal* listed new developments in neuroscience. One experiment funded by the Pentagon's research arm attempts to create headsets for wireless brain-to-brain communication. Elon Musk's company Neuralink has unveiled a device to control some human movements with a computer. Brain-computer interfaces (BCIs) are beginning to be medically successful. At Rice University, light decodes activity in one brain and links two brains together by magnetic fields. Surgically implanted devices strengthen memory by 37 percent. Energizing the brain with micro-pulses of electricity to boost learning abilities is being researched. A post-humanist world of chimeras is nearly here. Humans may be an intermediate step, and we will pass the baton from cellular to mechanical creatures.

## CRISPR

CRISPR will make it easy to direct our evolution. It is doubtful that we will be able to resist changing our germ cells, not just to cure diseases but to enhance individuals or groups. Replacing humans is a beguiling prospect because of our inequality, greed, cruelty, and drive to sack the planet for its energy. The hazard is that we will replace genes to form new class inequalities because of our competitiveness. The genetic code is so complex that we will make serious mistakes and end up in strange places. Artificial intelligence's ability to predict the consequences may temper our enthusiasm for changing our nature.

The crises springing from the redefinition of humans by the development of artificial intelligence, robots, chimeras, and changes to the genetic structure that can shape our species' evolution are accompanied by several other potential severe disruptions.

Humans are jammed together as our population approaches ten billion. Compaction intensifies interaction, magnifies stress, and escalates emotional temperature and pressure. The climate change crisis further increases the likelihood of conflicts because of large-scale migration. There may be a slowdown in population growth; however, migration because of climate change will likely exacerbate the effects of the increase.

There has been a suggestion that one-third of the planet's land surface should be allocated to rewilding. On the face of it, that seems incredibly unlikely; however, when the population reaches ten billion three-quarters of the way through this century, it will be followed by a decrease. As the world becomes more secular and women gain more access to work, they will not want to have children except as a lifestyle choice. A decrease in childbearing can alter the population exceptionally quickly and make the idea of bailing out the environment much more credible. Since women making money and freed of childbearing will demand more goods, entropy will certainly not decrease even if the population does.

Even now, as ecological conflict increases, our weaponry becomes more destructive. This danger can only increase. Our destructive capacities have increased exponentially with cyber warfare, small nuclear devices, long-distance unmanned armaments, and the weaponization of space. So far we have largely resisted using nuclear devices, which might give us hope that the human species will be able to handle new responsibilities. On the other hand, in 2022–23, threats of nuclear war have been made. It is no longer unthinkable.

Because we live within the illusion that we are central to all existence, we do not believe we will destroy our planet or ourselves. We are sure that we will always be indestructible and privileged at the center. Our belief in God and another world, our specialness to Him, and His plans for our salvation mean we cannot be destroyed. However, a world nuclear war is likely within this century, possibly soon.

Six elements will change the future for Homo sapiens: nuclear bombs, quantum computers, awesome artificial intelligence, having the means to alter our evolution, combining humans with machines, and climate change, which will increase conflicts and cause mass migrations. Is humanity equal to these responsibilities? I think not.

## IS HELP ON THE WAY?

There will soon be computers that process information more accurately and create predictive stories about the actions of nature and humans better than humans can. Combining artificial intelligence and technological innovations with the brain and nervous system to enhance human functioning is underway.

Artificial intelligence could assist us in going against nature to maximize human safety in a universe of entropy. As a cellular system constructed to enhance entropy and iron out gradients, it is challenging for us at present to think or act in a non-cellular way and prioritize safety over entropy.

Nevertheless, we must try to reduce our pain and hazard. The cellular acting in a non-cellular way is not on without some external assistance. By ourselves, living within our nature and gaining a peaceful life is a total contradiction. In the natural course of things, humanity will go off the cliff. These new developments may help us save ourselves if we find the wisdom to use them and cooperate with them.

Changes in the development of the human species offer a narrow opportunity to create a more rational and objective species that can measure itself not according to its myths but to an accurate understanding of how matter, including life, works. Anyone reading this recognizes that a far greater likelihood is that the changes will be governed by greed and cruelty and the maintenance of the mythical identities of smaller groups and the species. Humans are not noted for their rationality in their everyday life. They live by their stories and prefer them to the truth. However, even if the path is very narrow, some of these changes will be used to build us a partial refuge from entropy.

Psychotherapy for individuals will often use personality-modifying drugs combined with cognitive behavioral therapy, which attempts to replace distorted thinking with reality-based cognition. If we look at the species therapeutically, then artificial intelligence begins to look like a cognitive behavioral therapy for the species. CRISPR, along with the development of chimeras, looks like the personality modifications already being used with individuals.

# Chapter 14

## A Narrow Path to Safety

The Greek myth of Sisyphus is a metaphor for life lacking meaning. The gods condemned Sisyphus to spend eternity attempting to roll a stone up a mountain but dictated that it would roll down as he neared the top. He walked down the mountain for eternity, grabbed the stone and rolled it back up again without ever reaching the top, never concluding his struggle. In his book, *The Myth of Sisyphus*, Camus used the story as a metaphor for the meaninglessness of human life and then proceeded to explore the story's significance for humans trying to live a meaningful life in a meaningless universe. He concluded that the life of Sisyphus was redeemed by his contemplations on the way down the mountain, summarizing his life and effort and making it meaningful to himself. Sisyphus reflectively created his narrative.

We believe our nature and destiny is to create by gathering potential energy into complex matter. We believe we are successful even as our efforts repeatedly fall apart. We live like Sisyphus. In our careers, we describe ourselves moving up the ladder and coming down. We speak of getting to the top, succeeding over the rest, reaching the heights, losing our grip on the top, and bottoming out.

The Sisyphus myth is a scientific reality and not just a metaphor. The story presents us with the reality of the alternating concentration of potential energy rolling the stone up the mountain, and kinetic energy as it inevitably bumps along down the mountain to the bottom. Energy seeks the lowest energy state at the bottom rather than the highest energy state of concentration at the top. That is scientific fact and not just the prescient Greek myth. All energy in the cosmos finds the lowest energy state. All of life, as well as the cosmic process, is Sisyphean.

Is that all there is? The short answer is yes. Our role in the conversion from matter to energy is our meaning. How can we ameliorate our condition and find meaningful goals relative to our species' circumstances?

We suffer from evil and pain while the human and natural world threatens to break. We flirt with nuclear war and dangerously heat the climate. Our emphasis on consumption and consumerism facilitates the destruction of complex natural structures.

We attempt to stitch the pieces together — the world and ourselves. The dressings for our wounds include stories, claims of exceptionalism, and arguments about life's meaning. They ameliorate but do not heal. In comparison to nature's complexity, our tried-and-true human cleverness is trivial.

How can we moderate our direction when we fear losing our primary cultural identities: tribes, languages, and cultural traditions? The prospect of a single culture fueled by the worldwide dissemination of ideas and information and the globalization of goods and styles liberates some and threatens others. Many fear losing their spiritual world of high gods who conquer death. We deny our tragic circumstances. We dig in our heels and pretend that truths are lies. We become violent in defense of our illusions, which guarantee our survival — somehow, sometimes, in some way.

Science is the alternative, the bedrock for understanding reality. This book attempts to see what human life looks like if we set aside our species' subjectivity. Culture, politics, religion, economics, and capitalism are froth on the rivers of scientific discoveries.

There have been several attempts to confront the lack of meaning for our species and to get around the consequences of the laws of matter and energy. The title of Part Three is taken from a song fully realized by the songstress Peggy Lee. In the song, she reflects on seeing her house go up in flames, being distracted and entertained by a circus, falling in love, and losing it. To all these events, her response is, "Is that all there is?" even when she reflects on her life at its end. It is all disappointing, so break out the booze, and have a ball, if that's all there is.

In the song, matter, culture, love, and life are missing something. All are inadequate. The solution here is to focus on personal enjoyment and pleasure. The difficulty is it does not work, and the inherent destructiveness of the entropic arrow reduces us.

The Buddhist concept of impermanence is an early intuition of entropy as the central scientific reality. Entropy was intuitively placed at the center of human pain by Buddha. To paraphrase Buddha, "All things have parts and therefore come apart. Don't cling." Clinging causes pain. The constant dissolution of matter is the origin of pain. The suggested solution is to use the recognition of constant impermanence and cultivate the ability to detach oneself from the flux and flow of experience radically. Opt out.

## SO, WHAT TO DO?

Our biological structures and functions and the genetically constructed algorithms in our brains are imperatives. We must eat, kill, survive, reproduce, keep warm, couple, relate to our kin, trade information, form social groups, and belong. In all we do, we create more randomness than coherence. What we must do is, de facto, our meaning. We have no choice. It is what we do, as other species do what they must do. We serve entropy. That is the Sisyphean part of our existence. That is all there is, *but we seek a different conclusion*.

We add thoughts and images as we push the stone up the mountain. We dream of something more. We add ideas, philosophy, songs, stories, rhythm, heroic actions, freedom, a different life without the old necessities, and transcendent love. Our culture becomes our meaning. It makes our fate easier. We do that, but it does not overcome pain and entropy.

Robert Frost knew one thing: life just keeps on going. That is all there is — just the unfailing cosmic process. That is the meaning. Go with it.

In his novella *Candide*, Voltaire suggests that we abandon great ideas of life's significance, ambitions, competitiveness, status, grandiose actions, and struggles for dominance and wealth. He claims they only lead to conflict, manipulation, pain, and death. In the words of Candide, Voltaire suggests that we stop talking about all that and devote ourselves to modest living as we attempt to tend our garden peacefully. The trick here is that we all must agree.

Above all, we strive to be safe. Safety is the fundamental goal to assure continuance. We process information constantly about our internal and external environment to identify threats and opportunities. Ethics arise from natural law through the imperative of finding safety. *Ought* is a consequence of *is*.

Continuance is everything. Humans are a chemical process grounded in the laws of physics, including the second law of thermodynamics. We transform kinetic energy into chemical energy and back into kinetic energy through muscular movement and sound generation. This totality of processes continues until it is stopped. We do everything to maintain our life drive, even in hopeless and painful situations. Our task however is the creation of entropy.

A sliding scale exists between maintaining safety on the one hand and pursuing consumption with maximum intensity on the other. Maintaining long-term safety requires cooperation, whereas intense consumption is individually motivated. Primarily focusing on consumption, we destroy other cellular creatures and the environment. It can happen in the shortest time. Our actions accumulate potential energy, which bursts as kinetic energy to take energy from others by force or trickery. Accumulated potential energy includes money, power, and power-driven ideologies. Being unsafe, we become aggressive and increase competition and conflict.

## CREATING SAFETY

Morality is necessary to maintain our collective agreements to prevent harm. Individual and family safety involves "Safety for Everyone" (SAFE), starting with the most vulnerable. All of us are fragile, and any of us could be helpless or vulnerable — and probably will. We are only safe if the vulnerable are safe.

Safety for one and all is an imperative of our need to survive. We destroy, and we will be destroyed. I argue for an ethic associated with religion and philosophy where small groups of individuals devote some portion of the resources to aid the safety of others who are at risk.

These values go back to the Axial Age. The entropic demand for unrestricted consumption — leading to competition rather than cooperation and the development of hierarchies by individuals gathering all kinetic energy into potential energy for themselves and their families — has constantly undermined cooperative and communal values.

Let us look at the so-called axial values in historically significant religions. The increased focus on productivity that came about with intensive farming had as by-products hierarchy, nepotism, extreme wealth inequality, exploitation, forced labor including slavery, and systematic brutality.

Safety within the group declined radically compared to the substantially more equal relationships within the small bands of hunter-gatherers.

In resistance to this, different cultures expressed new values at about the same time. They advocated equal rights and opportunity, representative forms of government, the sanctity of the rule of law with universal inclusion, and limits to arbitrary and excessive use of power. Unfortunately, whenever these values were institutionalized, considerable gaps developed between the values of equality and the institutional practices that maintained hierarchy and prioritized executive power within religious institutions. The tension between the interior experience of faith and the conformity to public rituals became endemic, as it still is.

Zoroastrianism, founded in the mountain foothills of Persia, emphasized the care of children, hospitality to strangers, and espousing gratefulness and generosity. One sect among them advocated the equal sharing of wealth and property. At about the same time, the values of fairness, impartiality in applying the law, meritocracy, and personal cultivation of virtue were advocated in north China.

Confucianism, which developed later in China, focused on what makes a good society. Fundamental tenets included the belief in everybody's capacity to develop morally through reflections on one's behavior, followed by concerted attempts at improvement. The core belief was that the practice would lead to a harmonious society if individuals endeavored to perfect themselves. Moral cultivation was accomplished altruistically to perfect society, not for a reward.

Buddhism advocated living a simple, humble life without discomfort. Mahayana Buddhism emphasizes selfless love. Devotees should work for the well-being and enlightenment of their neighbors. The three negative states are greed, hatred, and delusion or ignorance. Buddhism is egalitarian religiously in that every human being has the potential to become Buddha, but it does not extend this equality to sectarian societies. Buddhism promotes nonviolence.

In Christianity, God rewards acts born out of love for one's neighbor. Jesus criticized the priestly class and favored the poor and the socially despised. He wanted to create a community that included pure people, excluding the corrupt, powerful, and wealthy. In the early community of solidarity, there was spiritual equality regardless of class or gender. The idea of mutual support continued over time but eventually did not include social and gender equality.

Islam advocated protection for the weak. It did not allow debt slavery or the enslavement of Muslims and non-Muslims living in Muslim countries. One should be honest, fair, and modest, give to the poor, spend with moderation, speak kindly, and be generous. All men are considered equal. There are fewer roles and lesser rights for women.

In philosophy, there is an emphasis on controlling the behaviors and attitudes that lead to pain, destruction, and self-destruction. Examples are Plato's *Republic*, Stoicism, Kant's moral law, Kierkegaard's writing on honesty and faith in everyday life, and the parables of Tolstoy.

The principle that everyone should be treated equally under the law, and that no one should be treated arbitrarily, exemplifies the Safety for Everyone principle. It is the finest human achievement, although, unfortunately, it is constantly subverted. Secondly, there is the achievement of representative government where everybody has a vote. Education for all is a landmark achievement for equality. However, particularly at advanced levels of education, there is frequently privileged access for the wealthy and discrimination against the poor and women. These principles reinforce equality and are significant achievements of economic and political organizations, even though they are commonly subverted.

Religion has embraced the kind and ethical treatment of others to grant them safety and preserve their integrity as God-given rights. However, there has also been suppression and subversion within religions as they organized to foster predatory elites. The laity has colluded in establishing the hierarchy in the belief that rituals, narratives, and supernatural powers bring them safety. Morality and communal welfare take a backseat. A similar subversion happened in democracy when autocracy was chosen because the people believed they would be safer with a strong leader rather than with their deliberations.

It is hard to imagine a world in which firmly rooted cellular behavior, which goes back to the beginning, could change to privilege the safe endurance of humans. How can we slow down and spread out the burning up and burning down of organized matter, including other cellular beings and especially our species? Instead of going for broke through competition and super-intensity, how could we tweak our behaviors toward prudent cooperation in the management of energy with an eye to the safety of all the constituents of the process? Finding a solution is way beyond my abilities. All I can do is show you the problem.

We require an alteration in our point of view if we are to find a modicum of safety. By relocating ourselves to the scientific reality of our existence we can achieve safety and give real meaning to our lives. Viewing ourselves enclosed in nature can be a container for our hopes. A change in viewpoint could change our reality, our moral presence with each other, and the institutions we establish to create safety for us, our families, and our neighbors. If we can cut down our grandiose preoccupation with exceptionalism and potential energy accumulation, then we will automatically focus on the moral intention to bring others safety and kindness. This could provide us with sufficient meaning for our lives in the context of our species.

Eleven thousand years since the transition to the Neolithic is not enough time for the human species to develop more than the most fundamental values. Considering that some of our ideals, ethical philosophies, political constitutions, and religions express similar values, a movement toward small secular and altruistic groups could, through cooperation, make the extension of Safety for Everyone their dominant desire. The example of a few can sometimes precipitate a moral conversion of many. This is barely possible; barely, barely possible. Since we have no importance within the universe, why not make the best of it? Let us make life safe and minimize pain. Let us have moderate expectations and give up potential energy accumulation and the explosive creation of entropy — if possible.

In the beginning — in the random way that is the backbone of evolution — on a tiny rock, a chemical process self-organized to break down molecules of matter and harvest energy. Later, some molecules of nucleic acids had an affinity in their structures that allowed them to hook up with each other along a backbone of sugars and phosphates. They wriggled about and attached themselves to identical elements, so the string duplicated itself. These processes of breaking down matter into energy to bring equilibrium into gradients, along with the assembly of molecules able to convey information on how to build little cellular machines, both became enclosed within a membrane with water from their initial environment.

Mitochondrion bacteria process matter into molecules that bind and release electrons to create energy efficiently. Over time, stimulated by environmental changes, the chemical processes became more efficient at processing energy in the making and breaking of molecular bonds that liberated electrons. Efficiency increased when the atmosphere changed to include high levels of oxygen. They formed the basis of all animal life when absorbed into another cell.

The evolutionary process arrived at two modes of efficiently processing energy. The first mode of behavior emphasized individual accumulations of matter and energy as potential energy that could be discharged into the environment in creative and destructive ways as the concentration reaches a lower energy state. Most creatures entirely focus on their singular life process. However, some creatures cooperated to support each other in their predatory process and to create safety by collectively resisting being consumed. For creatures in this mode, there is the problematic accumulation of energy, and then there is the painful, desperately resisted dissolution of the hoard — eventually, the sack breaks.

The cosmic arc provides humans with two paths: the accumulation of potential energy followed by spilling it intensely, and the equitable sharing of the resources for energy with the mutual granting of safety. The latter road conforms with energy processing according to the second law. It is superior to accumulating potential energy and spilling it intensely, but that is a matter for the universe to say, not for me. Both are chosen. Our choices can influence it. The reason for choosing cooperation and safety and equitably spreading out the process is to ameliorate individual pain.

At present the accumulation of potential and guarding it, on the model of what we do to preserve our bodies in the face of entropy, has become the selected paradigm for the organization of society economically and politically. The 6,000 years since the rise of proto-cities is too short a time for the selection to have been genetic; therefore, presumably, it could be reversed. The equitable distribution of energy concentrated as potentials that enable work and creativity to be distributed would be morally just, create more safety, and be more productive.

Humans are virtual particles, waves of possibility. Our existence is tragic as we perform the part created for us as chemical creatures while we fall apart and dissolve. We live in a storm of crackling energy seeking its ground, and we try to find a relatively safe harbor with some modicum of quiet. We can do something within the goals of nature, which would make our task more manageable. That would be sufficient.

Is that all there is? Yes, that is all there is.

# Appendix A

## Truth, Beauty, and Goodness

### TRUTH

Truth is based on the ability to predict how relationships take place. There is an unexplained ability of the language of mathematics to theoretically formulate the relationships between forces in the universe. The descriptions are borne out by close examination and repetitive testing that gives statistically high probability, often approaching certainty, that forces work in a particular way.

Our universe begins when the Higgs field gives mass to some of the excitations — particles — of other fields that are relativistic, thereby breaking symmetry and creating matter by restraining particles and energy within neutrons and protons.

The subsequent history of the universe is the breakdown of the potential energy held within composite matter into the random radiation of relativistic energy. At the same time the concentrations of temperature in composite matter compared to the low temperature of the vacuum in the universe is evened out. The universe processes to the lowest energy. In our human terms, where stories demonstrate a denouement revealing purpose and meaning, this seemingly arbitrary coming together and falling apart seems meaningless. In a human story it is as if a mistake was made and is being corrected.

Similarly, life, including our species, rests on the chemical mechanism in the cell which evolved to break down matter into radiation and destroy the unevenness of densities. This is our purpose. It is what we are good at, and why we exist. Our cells continue to do this until they themselves are taken apart.

Our species overvalues the putting together of composite matter because that is what we struggle to do to continue to exist and perform our function of breaking down matter. We have a fantasy that we are a special case within the universe, a conclusion that we base on the exaggerated importance given to our ability to put things together. We exaggerate because we do not want to see that our actual purpose and eminent ability is in taking things apart.

All this kind of sounds crazy, but it is what we do that makes our species' enterprise tragic.


## BEAUTY

Beauty exists in nature as low entropy. Less information is preferred. Shape, color, and movement simply expressed with few complications are the substance of beauty. Examples are the song of a bird against the background of silence, sunrises and sunsets scattering photons through the particles of the atmosphere and delineating clouds, the shape of the sun appearing to set in or rise from the sea, a tree standing alone, a dancer with force and acceleration extending himself as he rises in the air, a highly contoured human body, a full moon sitting alone in the sky particularly when it is closest to the Earth, the shape of a Siamese cat highlighted against a window, a single rose, a simple equation that holds within it the explanation of diverse phenomena.

Beauty is the antithesis of entropy but, as it is said, beauty is fleeting. Beauty is to entropy as potential energy is to kinetic energy.

Beauty is performative. It stands out and surprises. It does not stir emotion as much as awe. It gives striking evidence to a form of truth. Art also gives evidence of truth performatively, but the purpose of art is not beauty. Art holds in tension gradients, which it seeks to reconcile. Art is complex.

## GOODNESS

Goodness is a concept that is grounded in physics but is applicable only within human societies. In this respect, it is unlike truth and beauty. Goodness is the ability to move energy from potential to kinetic in a graduated way without creating chaos by enabling kinetic energy to create potentials equitably among the members of a human group. On the individual level it is the ability to transfer energy from one person to the other to enhance their potential energy. The transfer of energy is transparent, unambiguous, and does not permit manipulation.

# Appendix B

## Fabulation

Homo fabulator: We survive and fabulate. Time is driven by entropy, which is tearing down order and matter. We interpret time as progressing, as our building up according to our purposes.

We are rare epiphenomena of quantum fields. At our core we are not human. In the context of this large world, and in the context of the basic wavelength nature of fundamental energy movements, as a species we seem not only tiny but also an object — a construct. In fact, as a chemical process, we are constructs. This is what scares us about artificial intelligence.

What to make then of our beliefs? Every species communicates and develops habits of relating to each other and to their environment. Although we are more complex, we are all fundamentally doing the same thing.

Our responses to environmental stress become habitual and are copied throughout a culture. The actions are chosen subconsciously without free will. However, we disguise our lack of freedom by fabricating stories as to why we are acting in particular ways. Meaning and purpose to our life exist in the stories and only in the stories.

The difference in this book from most stories is that it assesses the implications of science for the beliefs and conduct of the human species. It is grounded on the axioms, hypotheses, and experiments of science, which describe the forces that create human relationships and institutions. Nevertheless, expressed in words rather than mathematics, this book is an interpretation — not a proof.